Springer Optimization and Its Applications

VOLUME 74

Managing Editor
Panos M. Pardalos (University of Florida)

Editor- Combinatorial Optimization
Ding-Zhu Du (University of Texas at Dallas)

Advisory Board
J. Birge (University of Chicago)
C.A. Floudas (Princeton University)
F. Giannessi (University of Pisa)
H.D. Sherali (Virginia Polytechnic and State University)
T. Terlaky (Lehigh University)
Y. Ye (Stanford University)

Aims and Scope
Optimization has been expanding in all directions at an astonishing rate during the last few decades. New algorithmic and theoretical techniques have been developed, the diffusion into other disciplines has proceeded at a rapid pace, and our knowledge of all aspects of the field has grown even more profound. At the same time, one of the most striking trends in optimization is the constantly increasing emphasis on the interdisciplinary nature of the field. Optimization has been a basic tool in all areas of applied mathematics, engineering, medicine, economics, and other sciences.

The series *Springer Optimization and Its Applications* publishes undergraduate and graduate textbooks, monographs and state-of-the-art expository work that focus on algorithms for solving optimization problems and also study applications involving such problems. Some of the topics covered include nonlinear optimization (convex and nonconvex), network flow problems, stochastic optimization, optimal control, discrete optimization, multiobjective programming, description of software packages, approximation techniques and heuristic approaches.

For further volumes:
www.springer.com/series/7393

Panos M. Pardalos · Pando G. Georgiev
Petraq Papajorgji · Britta Neugaard
Editors

Systems Analysis Tools for Better Health Care Delivery

Editors
Panos M. Pardalos
Department of Industrial and Systems
 Engineering
University of Florida
Gainesville, FL, USA

Petraq Papajorgji
Ministry of Education and Sciences
National Agency for Exams
Tirana, Albania

Pando G. Georgiev
Center for Applied Optimization
Department of Industrial and Systems
 Engineering
University of Florida
Gainesville, FL, USA

Britta Neugaard
James A. Haley VA Medical Center
Tampa, FL, USA

ISSN 1931-6828
ISBN 978-1-4614-5093-1   ISBN 978-1-4614-5094-8 (eBook)
DOI 10.1007/978-1-4614-5094-8
Springer New York Heidelberg Dordrecht London

Library of Congress Control Number: 2012953471

Mathematics Subject Classification (2010): 11T71, 14G50, 34L25, 34A30, 34A34, 37L30, 37N40, 41A05, 41A20, 41A21, 42A16, 51M04, 51N20, 60G25, 60J10, 60K15, 65C20, 65C50, 65D18, 65S05, 65T60, 68P25, 68P30, 68U20, 74J20, 78A46, 81U40, 90B10, 90B50, 90C11, 90C29, 90C90, 94A05, 94A15, 94A60, 97R30, 97R60, 97U70

© Springer Science+Business Media New York 2013
This work is subject to copyright. All rights are reserved by the Publisher, whether the whole or part of the material is concerned, specifically the rights of translation, reprinting, reuse of illustrations, recitation, broadcasting, reproduction on microfilms or in any other physical way, and transmission or information storage and retrieval, electronic adaptation, computer software, or by similar or dissimilar methodology now known or hereafter developed. Exempted from this legal reservation are brief excerpts in connection with reviews or scholarly analysis or material supplied specifically for the purpose of being entered and executed on a computer system, for exclusive use by the purchaser of the work. Duplication of this publication or parts thereof is permitted only under the provisions of the Copyright Law of the Publisher's location, in its current version, and permission for use must always be obtained from Springer. Permissions for use may be obtained through RightsLink at the Copyright Clearance Center. Violations are liable to prosecution under the respective Copyright Law.
The use of general descriptive names, registered names, trademarks, service marks, etc. in this publication does not imply, even in the absence of a specific statement, that such names are exempt from the relevant protective laws and regulations and therefore free for general use.
While the advice and information in this book are believed to be true and accurate at the date of publication, neither the authors nor the editors nor the publisher can accept any legal responsibility for any errors or omissions that may be made. The publisher makes no warranty, express or implied, with respect to the material contained herein.

Printed on acid-free paper

Springer is part of Springer Science+Business Media (www.springer.com)

Preface

This book presents some recent systems engineering and mathematical tools for health care along with their real-world applications by health care practitioners and engineers. Advanced approaches, tools, and algorithms used in operating room scheduling and patient flow are covered. State-of-the-art results from applications of data mining, business process modeling, and simulation in health care, together with optimization methods, form the core of the book. It illustrates the increased need of partnership between engineers and health care professionals.

In what follows, we present a brief outline of the contributed papers in this volume, which are collected in an alphabetical order of the contributors.

In Chap. 1, Dionne M. Aleman, Hamid R. Ghaffari, Velibor V. Mišic, Michael B. Sharpe, Mark Ruschin, and David A. Jaffray present a semi-infinite linear programming approach to solve high-resolution, convex quadratic optimization treatment problems in a reasonable amount of time. They also devise several computational improvements to the commonly used projected gradient algorithm that provide significant time savings when optimizations must be performed iteratively. Their approaches allow previously unwieldy treatment planning problems to be solved in a clinically viable amount of time.

In Chap. 2, Nebil Buyurgan and Nabil Lehlou present a study on the analysis of portable asset management strategies in hospitals. The problem addressed here is the unavailability of the portable assets when they are needed due to lost or hoarding, which lead to significant amount of staff time for search and underutilization of the assets. A simulation-based decision support tool is constructed to analyze the different processes and the impact of Radio Frequency Identification (RFID) technology on the widely adopted portable asset management models. The results suggest that the substantial gain could be realized by implementing RFID systems.

In Chap. 3, Camilo Mancilla and Robert H. Storer presents several data mining tools that can be used to investigate health outcomes, and provides a sample analysis of health care data to demonstrate their use. The tools include market basket analysis, text analysis, and predictive modeling. These tools can investigate also cancer treatments. The need to analyze real data is particularly necessary with the increased prominence of comparative effectiveness analysis.

In Chap. 4, Anastasius Moumtzoglou and Anastasia Kastania review different ways that stochastic integer programming has been used to improve efficiency and efficacy in health care delivery. For the purpose of this study health care delivery is divided in two areas: resource allocation and operations. In each area the stochastic components are identified and the algorithms and solution techniques that have been proposed in the literature are described. Current challenges and open questions are stated.

In Chap. 5, Neng Fan, Syed Mujahid, Jicong Zhang, Pando Georgiev, Petraq Papajorgji, Ingrida Steponavice, Britta Neugaard, and Panos M. Pardalos present a survey on e-health management. Changes in health care delivery have become so widespread and numerous that the idea of e-health has become one of excitement and prediction rather than intervention. On the other hand, the endorsement of e-health is spreading slowly. Few companies focus on population-oriented e-health tools partly because of perceptions about the viability and capacity of the market. Moreover, developers of e-health resources are a highly diverse group with differing skills and resources while a common problem for developers is finding the balance between risk and outcome. On the other hand, e-health presents risks to patient health information that involve not only appropriate protocols but also laws, regulations, and appropriate safety culture. Breaches of network security and international viruses have elevated the public awareness of online information and computer security, although the overwhelming majority of security breaches do not directly involve health-related data. Finally, as we believe in the implications of the genetic components of disease, we expect a significant increase in the genetic information of clinical records. The future vision is mobile personalized e-health in a patient-centered and patient-safety context.

In Chap. 6, Patricia Cerrito. We use a binary integer programming model to formulate and solve a nurse scheduling problem (NSP) which maximally satisfies nurse preferences. In a practical application of a VA hospital, besides considering the scheduling of two types of nurses (registered nurses and licensed practical nurses), two other types of employees (nursing assistants and health care techs), one nurse manager and a clinical nurse leader are also included in our model. Most of these employees are working full-time. In addition, we distinguish the schedule of weekdays and weekends with different requirements and different preferences for employees. Besides the requirements for each shift, we consider requirements for specific employees in some shifts in practical situations. The seven shifts do not necessarily have the same length in our model. Vacation time of employees is also considered in our model. Thus, the requirements for nurse scheduling are complicated and the objective is to maximize the satisfaction of preferred schedules of all these employees, including both nurses and other staffs. The presented model is complex, but efficiently solvable, satisfying the set of requirements in a particular application in a VA hospital.

In Chap. 7, Jennifer A. Pazour and Russell D. Meller study the pharmaceutical supply chain from a pharmaceutical distributor to a patient. The authors make comparisons between a traditional distribution center and a hospital pharmacy and discuss the technologies used in both facilities, with special emphasis on the order-fulfillment process. The authors review analytical models for order-fulfillment technologies prevalent in pharmaceutical distribution, including models for A-Frame systems, carousel systems, picking machines, unit-dose repackaging technologies, and automated dispensing cabinets. Finally, the authors provide conclusions and future research directions.

In Chap. 8, Elina Rönnberg, Torbjörn Larsson, and Ann Bertilsson describe automatic scheduling of nurses. The intention of this chapter is to provide a piece of practical experience that can help bridge the gap between advanced method development and the use of automatic nurse scheduling in practice. The approach described here is to take account of a real-life problem with all its details, and to use a straightforward meta-heuristic in order to deliver automatically generated schedules. The contribution of this chapter is based on the result of two case studies, which will provide insights into real-world examples, including evaluation and feedback from the wards.

We wish to express our deepest appreciation to the above-named authors who contributed their papers for publication in this volume. In addition, we are also very thankful to Springer Publishing Company for their generous support for this publication.

Gainesville, FL, USA	Panos M. Pardalos
Gainesville, FL, USA	Pando G. Georgiev
Tirana, Albania	Petraq Papajorgji
Tampa, FL, USA	Britta Neugaard

Contents

1 **Optimization Methods for Large-Scale Radiotherapy Problems** 1
Dionne M. Aleman, Hamid R. Ghaffari, Velibor V. Mišić,
Michael B. Sharpe, Mark Ruschin, and David A. Jaffray

2 **Portable Asset Management in Hospitals** 21
Nebil Buyurgan and Nabil Lehlou

3 **Stochastic Integer Programming in Healthcare Delivery** 37
Camilo Mancilla and Robert H. Storer

4 **An Expository Discourse of E-Health** 49
Anastasius Moumtzoglou and Anastasia Kastania

5 **Nurse Scheduling Problem: An Integer Programming Model with a Practical Application** 65
Neng Fan, Syed Mujahid, Jicong Zhang, Pando Georgiev,
Petraq Papajorgji, Ingrida Steponavice, Britta Neugaard,
and Panos M. Pardalos

6 **Clinical Data Mining to Discover Optimal Treatment Patterns** 99
Patricia Cerrito

7 **Exploring the Parallels Between a Hospital Pharmacy and a Distribution Center** 131
Jennifer A. Pazour and Russell D. Meller

8 **Automatic Scheduling of Nurses: What Does It Take in Practice?** 151
Elina Rönnberg, Torbjörn Larsson, and Ann Bertilsson

Optimization Methods for Large-Scale Radiotherapy Problems

Dionne M. Aleman, Hamid R. Ghaffari, Velibor V. Mišić,
Michael B. Sharpe, Mark Ruschin, and David A. Jaffray

1 Introduction

Mathematical models have been widely applied to the problem of designing highly customized radiotherapy treatment plans [1], but these previous approaches have almost exclusively focused on relatively moderate-sized treatments. The treatments previously studied include site-specific (e.g., head-and-neck, breast, prostate) *intensity modulated radiation therapy* (IMRT)—a treatment modality that allows for each beam in the treatment to have a unique distribution of radiation in order to deliver highly accurate dose—and Gamma Knife® treatments for small targets in the brain. For large-scale treatments, such as *total body irradiation* or very high-resolution Gamma Knife® Perfexion™ treatments, the optimization methods previously employed are no longer viable.

The work of this author was supported in part by The Canada Foundation for Innovation.
The work of this author was supported in part by the Natural Sciences and Engineering Research Council Undergraduate Summer Research Award.

D.M. Aleman (✉) • H.R. Ghaffari • V.V. Mišić
Department of Mechanical and Industrial Engineering, University of Toronto,
5 King's College Road, Toronto, ON Canada M5S 3G8
e-mail: aleman@mie.utoronto.ca; ghaffari@mie.utoronto.ca; velibor.misic@utoronto.ca

M.B. Sharpe • D.A. Jaffray
Department of Radiation Oncology, Princess Margaret Hospital, University of Toronto,
610 University of Avenue, Toronto, ON Canada M5G 2M9
e-mail: michael.sharpe@rmp.uhn.on.ca; david.jaffray@rmp.uhn.on.ca

M. Ruschin
Department of Radiation Oncology, Odette Cancer Centre, University of Toronto,
2075 Bayview Ave, Toronto, ON Canada M4N 3M5
e-mail: mark.ruschin@sunnybrook.ca

We investigate optimization methods to address the computational difficulties present in large-scale radiation therapy treatment optimization. We specifically focus on improving total body irradiation using IMRT and applying IMRT optimization mathematical models to Gamma Knife® Perfexion™. We consider the radiotherapy optimization process to consist of two sub-problems, a common approach in IMRT literature (e.g., [2–6]). These problems must be solved sequentially rather than simultaneously in large-scale treatment problems due to the massive data requirements in storing the effects of radiation delivery configurations on the patient's tissues. The first problem is to determine the relative positions of the radiation beams with respect to the patient's body. Once those beam orientations are obtained, the general radiotherapy optimization model is formulated as

$$\begin{aligned}&\text{minimize} \quad \phi(\mathbf{z}) \quad &\text{(RT - OPT)}\\ &\text{subject to} \quad \mathbf{z} = d(\mathbf{x})\\ &\quad\quad\quad\quad\; \mathbf{x} \geq 0\end{aligned}$$

where $\mathbf{x}$ represents the radiation intensity of the variables within our control; $d(\mathbf{x})$ is a function relating radiation intensities $\mathbf{x}$ to delivered dose $\mathbf{z}$; and $\phi(\mathbf{z})$ is a quantitative measure of treatment plan quality, where smaller values correspond to better treatments.

Define z_{js} as the dose delivered to voxel j in structure $s \in S$, which has v_s voxels. A voxel is a cube used to discretize the patient's body. In our approach, $\phi(\mathbf{z})$ is comprised of convex quadratic penalties $F_s(z_{js})$ that weight the over- and underdosage of each voxel j in structure $s \in S$. Letting $(\cdot)_+$ represent $\max\{\cdot, 0\}$, the penalty function for a single voxel is

$$F_s(z_{js}) = \frac{1}{v_s}\left[\underline{w}_s\left(\underline{T}_s - z_{js}\right)_+^2 + \overline{w}_s\left(z_{js} - \overline{T}_s\right)_+^2\right]$$

where $\underline{w}_s$ is the weight assigned to penalize any dose received under $\underline{T}_s$, and $\overline{w}_s$ is the weight assigned to penalize any dose received over $\overline{T}_s$. In previous approaches using penalty-based objectives (e.g., [2, 3, 6]), the thresholds at which over- and underdosing are the same. Our formulation allows for unique values at which to penalize overdose and underdose, and therefore yields increased flexibility by providing for "sweet spots" of radiation for structures at which no penalty is assigned. Graphically, the penalty can be represented as shown in Fig. 1 for a given structure, where T-u and T-o indicate the under- and overdosage thresholds, respectively. The RT-OPT objective is then to minimize the total penalty:

$$\phi(z) = \sum_{s \in S} \sum_{j=1}^{v_s} F_s(z_{js}).$$

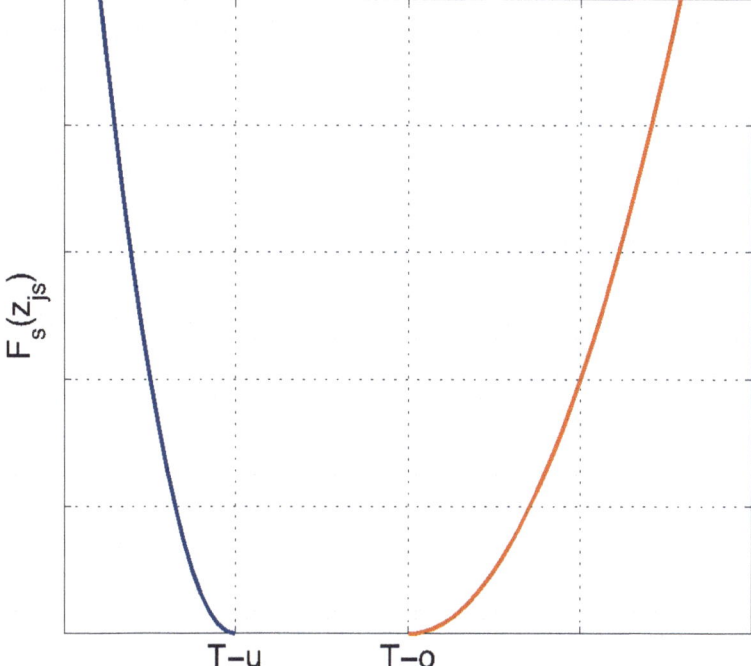

Fig. 1 Convex quadratic dose penalty with un-penalized sweet spot

2 Total Marrow Irradiation

Prior to receiving a bone marrow transplant, the patient's existing bone marrow must be eradicated in order to allow the donated stem cell transplant to successfully integrate with the body. One method of eliminating the bone marrow is through total body irradiation. Failure to destroy all the bone marrow will result in a transplant failure, so it is crucial to deliver enough radiation that the bone marrow is sufficiently eradicated.

The necessity of delivering high dose levels comes with more frequent and more severe toxic effects in healthy tissue [7], which can lead to a lower chance of a successful transplant. Further complications to the patient's health are caused by the method of radiation delivery. Because such a large area must be treated, clinicians typically place the patient far away from the isocenter—the central focus point of the beams of radiation—so that the area covered by each beam is large. The uncertainty in delivered dose results in the need to deliver high levels of radiation to ensure the bone marrow receives appropriate dose. It also prevents the use of conformal treatments that target just the bone marrow while sparing healthy organs.

Total marrow irradiation (TMI), a treatment that irradiates only the bone marrow while avoiding healthy tissues, can be achieved using IMRT. There has been limited research into the application of IMRT to TMI, mainly due to the computational issues present in designing complex treatments for such a large area. Using standard commercial planning systems, large reductions in dose to some organs can potentially be achieved [8, 9]. However, important organs such as the spinal cord are not considered. TMI has been considered using tomotherapy, and similarly shows that the dose organs can be significantly reduced [10, 11]. In contrast to these studies, we consider the TMI problem within the mathematical framework RT-OPT that we have successfully applied to TMI [12, 13]; we also consider non-coplanar beams (beams obtained from the movement of more than one linear accelerator component), which are necessary to deliver radiation to the patient at a standard 100 cm isocenter distance [14] so that analytical dose approximations are valid.

The two major subproblems in TMI are *beam orientation optimization* (BOO) and *fluence map optimization* (FMO). BOO determines the optimal beams from which to deliver radiation. Once these beams are obtained, FMO determines the optimal distribution of radiation for each beam. The distribution of radiation in each beam is delivered by considering each beam as being comprised of many smaller beamlets, each of which can deliver a radiation dose, called a *fluence*, independent of the other beamlets. The fluences for a set of beams are called a *fluence map*.

2.1 Beam Orientation Optimization

Beam orientation optimization has been well studied in the literature using a variety of approaches from genetic and evolutionary algorithms (e.g., [15–18]) to simulated annealing (e.g., [19–24]) to beam's-eye-view techniques (e.g., [21, 22, 25–29]) and more. Despite the large amount of research done in BOO, only a relatively small number of studies (e.g., [2, 3, 16, 18, 30–32]) have used the optimal fluence maps resulting from a set of beams to inform the selection beam orientations to use in the treatment plan. This simplification is largely due to the computational difficulties associated with having the FMO problem as the objective function of the BOO problem. These difficulties are highlighted by the fact that mixed integer approaches to combined BOO and FMO can only be performed if the beam solution space is restricted to a very small candidate set [33, 34].

The computational difficulties in addressing BOO and FMO simultaneously are even more evident in TMI, where the patient is on the order of 10 times larger than in the previously studied site-specific treatments, and 10 times more beams with 100 times more beamlets per beam are required to deliver an accurate treatment. Further, the patient size requires the use of non-coplanar beams, which increases the beam solution space to the point where little previous research has been able to consider non-coplanar beams [3, 21, 23, 30, 35–37]. Of these, all but [3] considered only a handful of non-coplanar beams.

Because it is widely accepted that the optimal solution to the FMO problem presents the most relevant measure of a beam set's quality [17–20, 23, 24, 33, 34, 36–46], and because it is essential to deliver a high quality treatment plan in TMI (more so than in traditional site-specific treatments), we seek to combine BOO and FMO. However, the TMI treatment planning problem is more difficult to solve due to the large patient size and beam solution space, so an algorithm that can move quickly through the solution space is desirable.

Thus, in order to solve the BOO problem, we employ the Add/Drop neighborhood search approach developed by [2] for coplanar beam selection. We extend the Add/Drop method to address the non-coplanar beam space necessary for TMI as described in [13]. In the Add/Drop method, each beam is analyzed in turn and replaced with an improving neighboring beam. Our enhancement redefines a beam's neighborhood as not simply a collection of all beams within a certain proximity, but only as nearby beams obtained from moving a single linear accelerator component. Therefore, each beam has multiple neighborhoods. The neighborhood examined in an iteration depends on historical improvements of that beam-component pair and probabilistic expectations of treatment plan improvement. Further details about the non-coplanar Add/Drop method are provided in [13].

Because each iteration of the Add/Drop method requires enumeration of the FMO solutions for each neighbor in the selected neighborhood, the computation time of the algorithm is dependent on the speed of the FMO optimization. Thus, we focus our efforts on improving the speed of the FMO optimization.

2.2 Fluence Map Optimization

As previously stated, IMRT optimization for TMI is more difficult than site-specific treatments due to the size of the patient. While a typical head-and-neck treatment contains ≈80,000 voxels, a TMI treatment contains ≈760,000 voxels. Empirically, our work has indicated that 30 beams are necessary to deliver a clinically acceptable TMI treatment, and each beam typically has ≈3,000 active beamlets. This results in 90,000 beamlet intensities that must be optimized in the fluence map optimization. The number of decision variables makes the use of Hessian-based algorithms, such as interior point methods that have been shown to yield optimal high-quality IMRT treatments [47], prohibitively expensive in terms of time and potentially numerically unstable.

We therefore apply a standard projected gradient algorithm with an Armijo line search [48, 49] to solve the FMO problem. Although projected gradient methods cannot guarantee an optimal solution, such methods are known to be fast and empirically return good solutions. Because the FMO formulation given in RT-OPT is a convex quadratic with only nonnegativity constraints, the local optimum approached by the projected method is the globally optimal solution. In practice, projected gradient methods have been shown to return quality treatment plans in

IMRT optimization [2, 3], even applied to TMI [12, 13]. However, while these solutions are obtained very quickly for site-specific treatments [47], about 45 min is required to return a solution for TMI.

Because the FMO must be performed for each neighbor in the Add/Drop algorithm used to select beam orientations, lengthy FMO computations significantly impact the solution space that can be searched in the fixed 12-h treatment planning limit imposed by clinicians. To speed up the projected gradient algorithm, we examine alternate line search techniques and a warm start approach.

2.2.1 Alternate Line Search Techniques

The projected gradient algorithm starts from a near-zero solution (in this case, a solution with all beamlets delivering almost zero intensity) and then picks a direction to move to obtain the maximum improvement in the objective function. This direction is the gradient of the objective function; specifically, since our FMO is a minimization problem, the direction is the negative gradient. Once this descent direction is obtained, the algorithm moves from the current point some distance in this direction. The determination of this distance is called a line search because the algorithm must search the descent direction for a distance to move, called a step length, so that the new solution yields at least some minimum amount of improvement in the objective function.

Most commonly, the majority of computation time spent in projected gradient methods is searching the descent direction for an appropriate distance to move the current solution. We explore three methods of performing the line search to determine which technique most improves the computation time of the algorithm.

First, we employ a traditional backtracking method [48]. In backtracking, we select some initial step length Δ, and if the solution located at that point provides insufficient objective function improvement, we decrease the step length by a factor δ and examine the resulting solution. This process is repeated until an appropriate step length is found.

After examining the step lengths taken by the traditional backtracking line search, we observed that for our specific optimization model, the final step lengths taken in the first few iterations were far larger than in most subsequent iterations. We therefore designed a modified backtracking line search where the initial step length is drastically reduced after a fixed number of iterations.

We also observed that the final step lengths taken in most iterations were generally very small. Yet, the backtracking algorithm was started with a fairly large value of Δ. So, the third line search method we attempted was a simple forward line search operating in the same manner as the backtracking line search, with the exception that we start with a very small step length and increase the step length by a factor of δ until a solution with sufficient objective function improvement is found.

Optimization Methods for Large-Scale Radiotherapy Problems

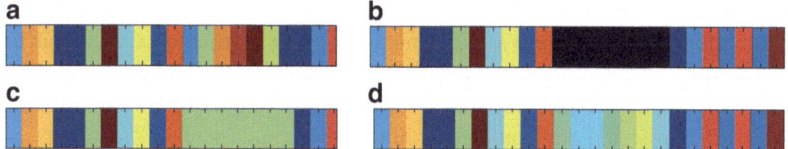

Fig. 2 Warm start illustration. (**a**) Initial solution. (**b**) Old beam replaced with new beamlets. (**c**) Average fluence initialization. (**d**) Least-squares fluence initialization

2.2.2 Warm Start Approaches

Although projected gradient methods start at near-zero solutions, the iterative process taken by the Add/Drop method supplies useful information about fluences of each beam in each iteration. Rather than start at a near-zero solution, we can use this information to start a near-optimal solution, which will significantly reduce the number of iterations needed to converge to a solution.

From one iteration to the next, only one beam changes in the Add/Drop method, meaning that only a subset of beamlet variables are new. Figure 2a illustrates beamlets values for some solution, which then become the values shown in Fig. 2b when one set of beamlets is replaced by new variables. The beam that changes is moved to a neighboring beam, which is near to the original beam. Therefore, it is likely that the fluences of the other beams in the solutions will be similar in the new solution to their values in the previous solutions.

Rather than discard this intuitive knowledge about the beamlet values from one iteration to the next, we use the previous optimal beamlet values as a warm start to the projected gradient algorithm. The beamlets from the old beam are replaced by new beamlets from the new beam, but, because these beamlets are new, we devise two methods of assigning their values in the initial projected gradient solution.

First, we initialize each of the new beamlets to have the average value of the beamlets of the old beam (Fig. 2c). This averaging is a computationally inexpensive method of exploiting the fact that the old beam and new beam will likely have similar overall intensity. Second, we initialize the values of the new beamlets to values that will most closely approximate the dose delivered by the original beamlets (Fig. 2d). This approximation is done using a least squares optimization.

If the new beam is the th beam in the solution ( ) and the set of beamlets in this beam is B with intensities $\tilde{x}_i$, then the least squares optimization is given by

$$\text{minimize} \quad \sum_{s \in S} \sum_{j=1}^{v_s} \left(\tilde{z}_{js}^{(\ )} - z_{js}^{(\ )} \right)^2 \quad \text{(LSQ)}$$

$$\text{subject to} \quad \tilde{z}_{js}^{(\ )} = \sum_{i \in B} D_{ijs} \tilde{x}_i \quad s \in S, j = 1,\ldots,v_s$$

$$\tilde{x}_i \geq 0 \quad i \in B$$

where $\tilde{z}_{js}^{(\ )}$ is the dose delivered by the new beamlets to voxel j in structure s and $z_{js}^{(\ )}$ is the dose delivered by the previous beamlets to voxel j in structure s. Although using a least squares optimization to initialize the new fluence values is potentially computationally intensive, it is possible that the reduction in the number of iterations performed by the projected gradient algorithm will still result in a faster solution.

2.3 Results

The algorithms were tested on a single TMI patient from The Princess Margaret Hospital (Toronto, ON, Canada). Each type of projected gradient method was executed on a Dell Intel Core 2 Duo with a 2.4 GHz CPU and 8 GB of RAM. Because the Add/Drop algorithm returns a locally optimal solution, the quality of the solution may be affected by the starting point. For robustness, we test the algorithms using 10 different randomly generated 30-beam solutions. The initial fluences of all of the beams were set to 0.3 Gy. Each variant of the projected gradient method terminated when successive iterations resulted in a relative objective function improvement of less than 0.01. For the reduced step projected gradient implementation, the step length was reduced after three iterations. For both the backtracking and reduced step implementations, the initial step length was 50 and $\delta = 0.25$. For the forward line search implementation, the initial step length was 3 and $\delta = 10$.

Table 1 illustrates the performance of each of the line search methods. In terms of the number of iterations and computation time, the reduced step variant performed the best, while the forward line search performed the worst. The poor computational speed of the forward line search results from the fact that, despite the small step size in a majority of iterations, the line searches with large step sizes required more computation time than what was gained by the small step sizes.

Somewhat surprisingly, the reduced step method resulted in the worst objective function values when used in the Add/Drop algorithm. One possibility for poor solution quality is that from our observations of the backtracking method, after the

Table 1 Comparison of the line search methods

		Backtracking	Reduced step	Forward
Add/drop iterations	Mean	16	13.7	16.1
	St. dev.	2.94	6.52	5.47
	Minimum	12	7	8
	Maximum	20	23	23
Time (min)	Mean	43.74	30.94	47.85
	St. dev.	9.04	15.87	16.41
	Minimum	31.70	13.40	22.70
	Maximum	57.30	51.70	68.10
FMO obj.fn. value	Mean	13,298.29	17,458.25	14,695.36
	St. dev.	2,062.89	4,657.02	2,750.92
	Minimum	11,822.90	12,230.20	11,724.50
	Maximum	18,492.00	23,452.30	19,184.00

first few iterations, there are infrequent iterations where the step size is large. But, the presence of those occasional large step sizes appears to be sufficient to guide the algorithm toward better solutions faster. Additionally, the Add/Drop algorithm terminates earlier in the reduced step implementation, resulting in a worse solution.

To assess treatment quality, *dose-volume histograms* (DVHs) are employed. This histogram is a measure of the cumulative dose received by a given structure, specifying the fraction of each structure's volume that receives at least a certain amount of dose. The DVH plots for healthy organs should fall as close to the origin as possible, while the DVH plot for the target structure should mimic a vertical line at the prescription dose.

Figure 3 shows the DVHs for the final solution obtained by the Add/Drop method for each line search variation. Because we consider 15 organs in addition

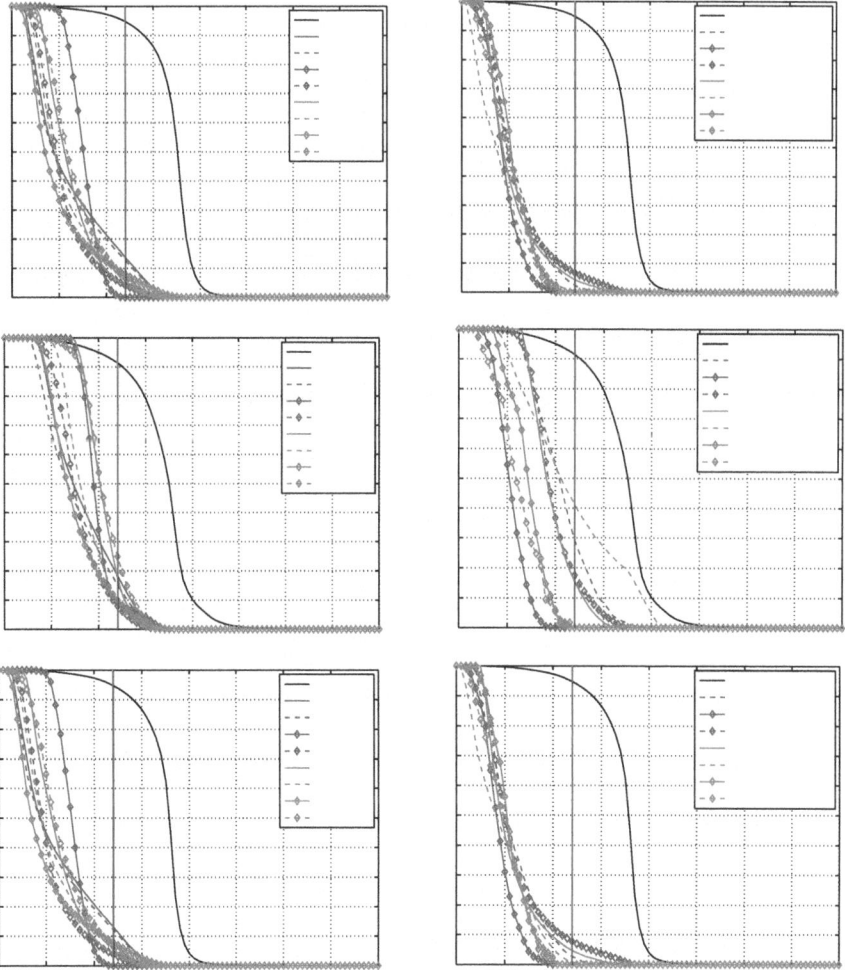

Fig. 3 DVHs from the line search methods, divided into two histograms per plan for clarity. *Top*: Backtracking. *Middle*: Reduced step. *Bottom*: Forward

Table 2 Comparison of the warm start methods

		Cold start	Warm start, averaging	Warm start, least-squares
Add/drop iterations	Mean	10.15	40.00	4,154
	St. dev.	2.44	6.58	4.45
	Minimum	7.00	31.00	36.00
	Maximum	15.00	55.00	52.00
Time (min)	Mean	70.88	18.17	17.28
	St. dev.	14.10	2.58	1.65
	Minimum	47.70	13.00	13.70
	Maximum	95.30	22.60	19.70
FMO obj.fn. value	Mean	14,906.08	12,016.72	11,572.92
	St. dev.	3,745.60	2,452.70	2,253.37
	Minimum	11,184.00	10,310.60	9,990.00
	Maximum	21,613.20	18,355.80	17,302.50

to the target structure (the bone marrow, labeled hemiPTV, where PTV stands for *planning tumor volume*), the organs for each treatment plan are displayed in two graphs, and the target is displayed in both for reference. The backtracking approach (Fig. 3, top) and the forward approach (Fig. 3, bottom) obtain nearly identical treatments, both of which meet clinical guidelines for target dose coverage. Because highly accurate TMI treatments have not been achievable in clinical settings, the healthy tissue responses to radiation in this scenario are unknown, so it is certain whether or not the organs are spared. Thus, the primary goal for the organs is to achieve a significant separation in the DVH curves, meaning that the organs receive far less dose than the bone marrow. The reduced step plan (Fig. 3, middle) is significantly worse than the other two plans, as predicted by the objective function values in Table 1.

Based on the line search results, the warm start approaches were each applied to the projected gradient algorithm with the backtracking line search. Table 2 shows the performance of the averaging and least-squares warm start approaches compared to no warm start (cold start). It is evident from these values that both warm start methods far outperform the cold start. The average FMO computation time is cut from 70 min down to about 18 min in both warm start methods, a 75% improvement. This time savings allows for four times as many Add/Drop iterations, meaning that four times the solution space is explored. As expected, the increased solution space exploration results in better BOO objective function values.

Figure 4 demonstrates that the quality of the treatment plans with warm starts is improved. Although the cold start solution is very good, the warm start solutions are able to obtain better target coverage without sacrificing any overdosage. Organ sparing between all three plans is clinically indistinguishable.

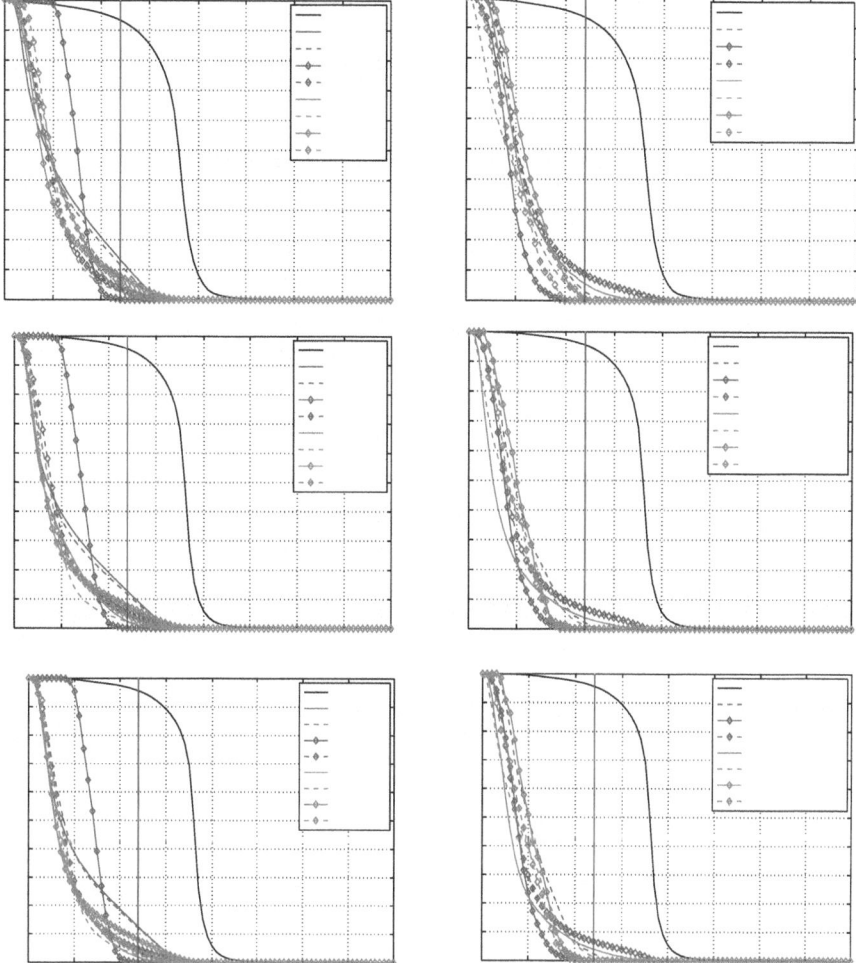

Fig. 4 DVHs from the warm start methods, divided into two histograms per plan for clarity. Top: Cold start. Middle: Warm start, averaging. Bottom: Warm start, least-squares

3 Gamma Knife® Perfexion™

The other large-scale radiotherapy treatment planning problem that we address is the design of treatments using Gamma Knife® (Elekta, Stockholm, Sweden) equipment. Gamma Knife® units are primarily used to treat brain tumors and lesions [50, 51], and are known for exceptionally high accuracy in radiation delivery obtained through the use of 60_{Co} beams [52] rather than photon beams used by linear accelerators. In Gamma Knife® treatment, photon beams are directed toward the patient from multiple *sectors* at several isocenters. Perfexion™, the newest generation of

Gamma Knife® machines, builds upon the older Gamma Knife® machines by allowing improved access to the cranial volume, variations in beam sizes, and, perhaps most importantly, automation of patient positioning and beam size modulation, allowing for highly customizable treatments to be delivered in a practical way in clinical settings.

Gamma Knife® treatment planning methods have included non-linear programming and non-linear mixed-integer programming [53, 54] to determine beam delivery, but these models are mathematically difficult to solve and require massive data storage, limiting the treatment size and voxel resolution that can be considered. *Isocenter optimization*, the optimization of isocenter (patient) locations has been approximated as the well-known ball-packing problem [55, 56] in previous studies [53, 54], but such an approach ignores Perfexion™'s capability of delivering non-convex treatment shapes.

The manual adjustments required in older Gamma Knife® treatments to change patient position and beam delivery meant that intricate treatments could not be delivered in a reasonable amount of time and effort, and so research efforts into optimization of Gamma Knife® treatments waned. Previous Gamma Knife® studies [53, 54, 57–60] are largely unable to incorporate delivery time and patient positioning in the optimization model in an efficient and practical way. Additionally, they do not address the new capabilities of Perfexion™ units. The optimization methods proposed here, adapted from [61] (currently the only mathematically-driven Perfexion™-specific treatment planning approach), account for the flexibility of Perfexion™, and can consider large-scale treatment sizes in a clinically viable amount of time.

3.1 Relation to IMRT Optimization

The amount of customizability in Perfexion™ precludes the use of previous Gamma Knife® optimization methods, but it results in a treatment optimization model that is similar to IMRT optimization, which has been well-studied [1]. Like IMRT optimization, Perfexion™ treatment planning can be segmented into two primary subproblems: isocenter optimization and *sector duration optimization* (SDO). Isocenter optimization is akin to beam orientation optimization in that the objective is to optimally locate the beams relative to the patient; in IMRT, the beams move around the patient, while in Gamma Knife®, the couch on which the patient is placed moves while the beams are fixed.

Sector duration optimization is the optimization of the amount of time each of eight sectors delivers radiation. In Perfexion™, each sector can independently deliver beams of diameter 4 mm, 8 mm, or 16 mm; SDO specifically determines the optimal time delivery of each sector at each possible beam size. This delivery is analogous to the FMO problem in IMRT, where the beam locations (isocenters) are fixed, and the amount of time each beamlet delivers radiation is optimized.

Because of the similarities between Perfexion™ optimization and IMRT optimization, we formulate the Perfexion™ treatment planning problem as an IMRT problem and attempt to apply IMRT optimization techniques. However, just as IMRT optimization methods break down in the face of large-scale TMI planning, they also break down for Perfexion™. Although Perfexion™ is used to treat small areas in the head, the accuracy of Perfexion™ means that the patient must discretized at a much higher level to match the precision of the 60_{Co} beams; voxel sizes are typically less than $1\,\text{mm} \times 1\,\text{mm} \times 1\,\text{mm}$. To model an entire head, this resolution results in over 100 million voxels in a typical $512 \times 512 \times 512$ voxel mask. Even though there are far fewer decision variables in Perfexion™ than in IMRT—8 sectors $\times$ 3 sizes $\times m$ isocenters $= 24m$ variables—the vast number of voxels makes optimization time consuming. The projected gradient method that was successful in TMI requires over 155 min to solve an SDO problem with 15 isocenters (360 variables).

Like BOO in IMRT, we consider the objective function of the isocenter optimization problem to be the optimal solution to the SDO problem using a given set of isocenters. But, due to the newness of Perfexion™, a robust and efficient SDO model and optimization method has not previously been developed. Thus, before we can address isocenter optimization, we focus on improving our SDO optimization technique so that the SDO optimization can be incorporated into isocenter optimization.

3.2 Sector Duration Optimization

The projected gradient algorithm requires a prohibitively long amount of time to solve the SDO problem, in large part due to the time needed to perform objective function calculations with so many voxels. Thus, we employ a semi-infinite linear programming (SILP) approach [62] that does not require objective function evaluations to solve the SDO problem.

An SILP is a linear program with an infinite number of constraints used to approximate a convex function, as shown in Fig. 5. Our convex quadratic SDO model, which follows directly from the general radiotherapy optimization problem given in RT-OPT is

$$\text{minimize} \quad \sum_{s \in S} \sum_{j=1}^{v_s} \frac{1}{v_s} \underline{w}_s \left(\underline{T}_s - z_{js} \right)_+^2 + \overline{w}_s \left(z_{js} - \overline{T}_s \right)_+^2 \quad \text{(SDO)}$$

$$\text{subject to} \quad \mathbf{z} = d(\mathbf{t})$$

$$t_{Ibc} \geq 0 \quad I \in \ , b \in B, c \in C$$

Here, decision variable t_{Ibc} is the amount of time to deliver radiation from sector $b \in B$ at size $c \in C$ to isocenter $I \in \Theta$, where the selected isocenters Θ are input into the model. To convert this model, which has a convex objective function and linear constraints, to a model suitable for SILP, we can transform SDO into a semi-conic (so-called due to the maximum functions in the constraints) programming problem simply by moving the objective function to the constraints:

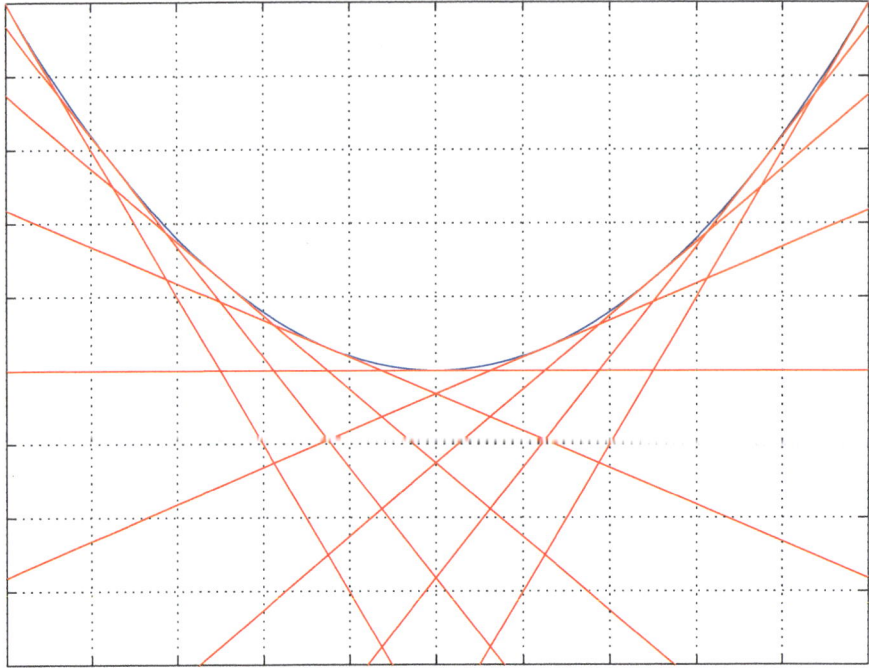

Fig. 5 Representation of transformation of a convex function to an SILP

$$\text{minimize} \quad \sum_{s \in S}(\,_s +\,_s) \quad \text{(SDO - SemiConic)}$$

$$\text{subject to} \quad \sqrt{\sum_{j=1}^{v_s} \frac{\overline{w}_s}{v_s}\left(z_{js} - \overline{T}_s\right)_+^2} \leq \,_s \quad s \in S$$

$$\sqrt{\sum_{j=1}^{v_s} \frac{\underline{w}_s}{v_s}\left(\underline{T}_s - z_{js}\right)_+^2} \leq \,_s \quad s \in S$$

$$\mathbf{z} = d(\mathbf{t})$$

$$t_{lbc} \geq 0 \quad\quad I \in \,, b \in B, c \in C$$

With SDO-SOCP, we can apply the SILP algorithm presented in [62]. The algorithm starts from an initial solution that is optimal to a reduced version of SDO-semiconic that considers some (but not necessarily all) of the constraints in the full model. Because only a subset of the constraints is considered at this point, some constraints from the full model will be violated. An interior point constraint generation method is used to identify these violated constraints. Some of these constraints are added to the reduced model, and the current solution point is moved enough so that it is feasible to the new larger reduced model. The algorithm then identifies a path from the relocated current point to the center of the feasible region of the new reduced

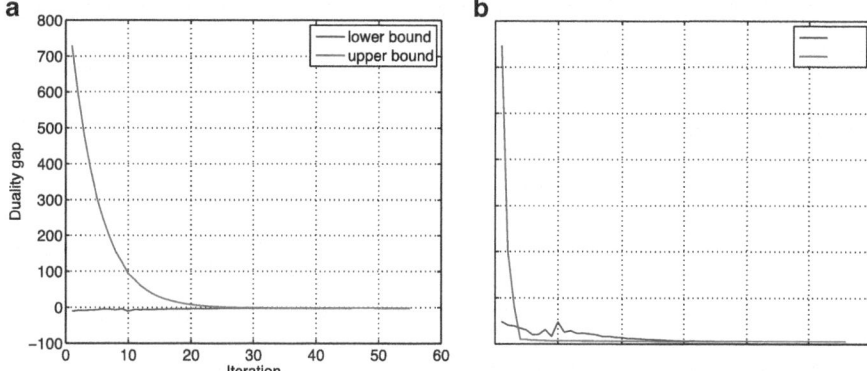

Fig. 6 SILP convergence. (**a**) SILP duality gap per iteration. (**b**) Comparison of projected gradient and SILP objective function values per iteration

model, and then moves to a point near this central path. At this new point, some constraints from the full model will be violated. Those constraints are added to the reduced model, and the process continues until the duality gap, which can be considered an upper bound on the distance from optimality, is sufficiently small. This algorithm is shown to obtain the optimal solution in a finite number of iterations [62].

To reduce unnecessary computations, only voxels within a reasonable vicinity of the target structures are considered. This simplification reduces the number of voxels from $512 \times 512 \times 512 = 134$ million to around 30,000.

3.3 Results

We tested the SDO approach using 100 randomly generated sets of isocenters. By testing on a large number of random isocenter locations, we can assess the ability of the SDO algorithm to deliver quality treatment plans regardless of patient position. The algorithm is run until the duality gap is smaller than 1e-5.

Figure 5a illustrates the duality gap per iteration obtained by the SILP approach. From this graph, it is apparent that SILP, although starting from a poor solution that is infeasible to most constraints, quickly converges to a near-optimal solution. Figure 5b also demonstrates that the SILP method starts from a very bad treatment, but within five iterations, finds a better solution than projected gradient. Note that the solutions in SILP are monotonically improving, while the projected gradient solutions do not exhibit this desirable property due to the zig–zag nature of projected gradient methods which may overshoot the optimal solution and have to reverse directions.

Figure 7 portrays the computation time required by the SILP algorithm. For 100 randomly generated instances of 15 isocenters, Fig. 7a shows that SILP far outperforms

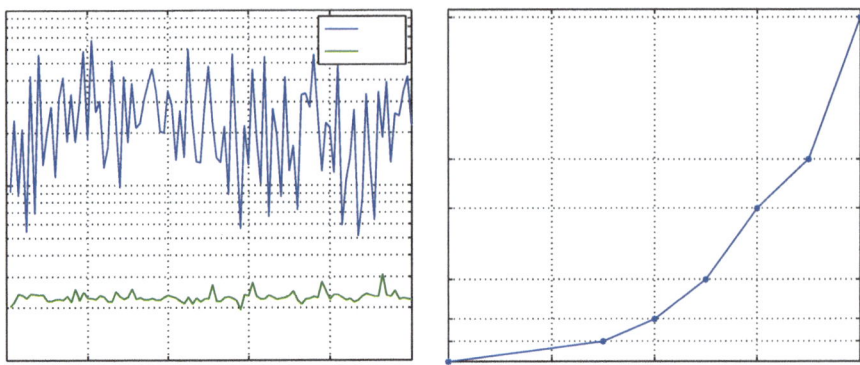

Fig. 7 SILP computational performance. (**a**) Comparison of projected gradient and SILP total computation time (min) over 100 instances of 15 isocenters. (**b**) Mean computation time (min) required by SILP for increasing numbers of isocenters

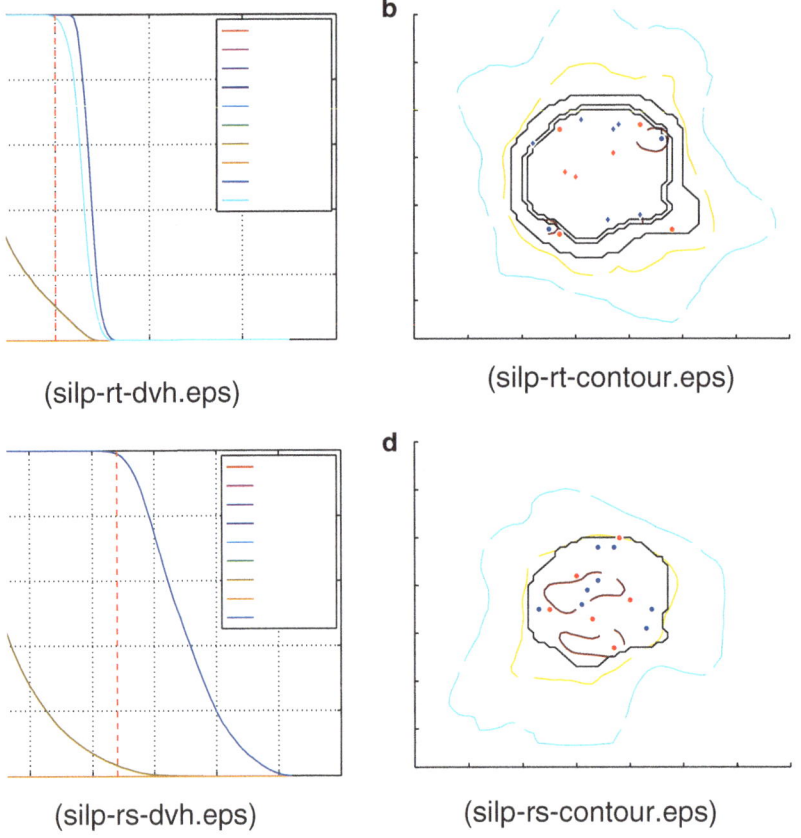

Fig. 8 SILP treatment plans. In the dose contour images, the structure on the *left* overlapping the targets is the brain stem. *Top*: Radiotherapy. *Bottom*: Radiosurgery

projected gradient, with the average SILP time of 22 min showing a factor of 7 improvement over the mean projected gradient time of 155 min. Furthermore, the SILP times are fairly consistent, while the projected gradient times are highly variable. As the number of isocenters increases, the computation time of SILP increases predictably (Fig. 7b). This consistency may help clinicians better plan their treatment design timelines.

As shown by Fig. 8, the SILP method delivers quality treatments for both radiotherapy and radiosurgery treatments. The difference between these two types of treatments is that in radiotherapy, the emphasis is on delivering a very uniform dose to the target, with little or no hotspots or coldspots; slight dosage outside the target is acceptable. In radiosurgery, the emphasis is on obtaining a highly conformal treatment, that is, almost no dose outside the target; cold spots must be avoided, but overdosage inside the target is acceptable up to 200% of the prescription dose. For radiotherapy, SILP was able to obtain a uniform dose to both the gross tumor volume (GTV) and the PTV, both of which have a prescription dose level of 2 Gy, without overdosing any targets, even though the brain stem is overlapping with part of the PTV (Fig. 8, top). For radiosurgery, SILP again spares all of the organs, and delivers a tightly conformal dose to the GTV (prescription dose of 12 Gy), as shown by the bottom dose contour in Fig. 8. Note that there is no PTV in radiosurgery.

4 Conclusions

Despite the difficulty in designing plans for large-scale radiotherapy treatments, it is possible to use mathematical models to optimize the quality of the treatments. With over 700,000 voxels and 90,000 beamlet decision variables in total marrow irradiation, our non-coplanar neighborhood search method for BOO [13] and projected gradient algorithm for FMO yielded high quality treatments in a clinically feasible 12-h time limit. When used in conjunction with an iterative BOO algorithm, improvements to the commonly used projected gradient method in terms of warm starts provided substantial gains in computation time and moderate gains in treatment plan quality. Alternate line search techniques did not result in significantly improved computation time, and even slightly worsened treatment plan quality.

For Perfexion™, the computational difficulties in calculating the objective function were mitigated by eliminating voxels positioned far from the targets, and by formulating the problem as a semi-infinite linear program. This formulation exploited the convex quadratic structure of the basic radiotherapy optimization model to obtain a semi-conic reformulation, which could then be solved using the interior point constraint generation algorithm described in [62]. By avoiding calculation of the objective function, computation times were reduced by a factor of 7 and the resulting treatment quality meets or exceeds clinical guidelines. The flexibility of the RT-OPT model allows for both radiotherapy and radiosurgery treatments to be developed within the same model and same algorithm with only minor modifications to the penalty weighting parameters.

Although there are numerous future directions that can potentially improve the quality of these approaches to radiotherapy, most notably incorporation of continuous radiation delivery as opposed to the current step-and-shoot delivery, our results show that it is possible to circumvent computational difficulties in radiotherapy optimization for very large treatment areas and very high voxel resolution.

References

1. Romeijn, H.E., Dempsey, J.F.: Intensity modulated radiation therapy treatment plan optimization. TOP 16(2), 215–243 (2008)
2. Aleman, D.M., Kumar, A., Ahuja, R.K., Romeijn, H.E., Dempsey, J.F.: Neighborhood search approaches to beam orientation optimization in intensity modulated radiation therapy treatment planning. J. Global Optim. 42(1), 587–607 (2008)
3. Aleman, D.M., Romeijn, H.E., Dempsey, J.F.: A response surface approach to beam orientation optimization in intensity modulated radiation therapy treatment planning. INFORMS; J. Comput.: Comput. Biol. Med. Appl. 21(1), 62–79 (2009)
4. Lim, G.J., Holder, A., Reese, J.: A clustering approach for optimizing beam angles in IMRT planning. In Proc IERC Annual Conf. IERC (2009)
5. Lim, G.J., Choi, J., Mohan, R.: Iterative solution methods for beam angle and fluence map optimization in intensity modulated radiation therapy planning. OR Spectrum 30, 289–309 (2008)
6. Romeijn, H.E., Ahuja, R.K., Dempsey, J.F., Kumar, A.: A new linear programming approach to radiation therapy treatment planning problems. Oper. Res. 54(2), 201–216 (2006)
7. Clift, R.A., Buckner, C.D., Appelbaum, F.R., Bryant, E., Bearman, S.I., Petersen, F.B., Fisher, L.D., Anasetti, C., Beatty, P., Bensinger, W.I., Doney, K., Hill, R.S., McDonald, G.B., Martin, P., Meyers, J., Sanders, J., Singer, J., Stewart, P., Sullivan, K.M., Witherspoon, R., Storb, R., Hansen, J.A., Thomas, E.D.: Allogeneic marrow transplantation in patients with chronic myeloid leukemia in the chronic phase: a randomized trial of two irradiation regimens. Blood 77, 1660–1665 (1991)
8. Aydogan, B., Mundt, A.J., Roeske, J.C.: Linac-based total marrow irradiation (IM-TMI). Technol. Cancer Res. T 5(5), 513–519 (2007)
9. Aydogan, B., Roeske, J.C.: Feasibility study of linac-modulated total marrow irradiation (IM-TMI). Med. Phys. 34(6), 2455 (2007)
10. Schultheiss, T.E., Wong, J., Liu, A., Olivera, G., Somlo, G.: Image-guided total marrow and total lymphatic irradiation using helical tomotherapy. Int. J. Radiat. Oncol. Biol. Phys. 67, 1259–1267 (2007)
11. Wong, J.Y.C., Liu, A., Schultheiss, T., Popplewell, L., Stein, A., Rosenthal, J., Essensten, M., Forman, S., Somlo, G.: Targeted total marrow irradiation using three-dimensional image-guided tomographic intensity-modulated radiation therapy: an alternative to standard total body irradiation. Biol. Blood Marrow Transplant. 12(3), 306–315 (2006)
12. Mišić, V.V., Aleman, D.M., Sharpe, M.B.: Total marrow irradiation using intensity modulated radiation therapy optimization. In Proc IERC Annual Conf. IERC, (2009)
13. Mišić, V.V., Aleman, D.M., Sharpe, M.B.: Neighborhood search approaches to non-coplanar beam orientation optimization for total marrow irradiation using IMRT. Eur. J. Oper. Res. 205(3), 522–527 (2010)
14. Khan F.M.: The Physics of Radiation Therapy. Lippincott William and Wilkins, (1994).
15. Ezzell, G.A.: Genetic and geometric optimization of three-dimensional radiation therapy treatment planning. Med. Phys. 23, 293–305 (1996)
16. Haas, O.C., Burnham, K.J., Mills, J.: Optimization of beam orientation in radiotherapy using planar geometry. Phys. Med. Biol. 43, 2179–2193 (1998)

17. Li, Y., Yao, J., Yao, D.: Automatic beam angle selection in IMRT planning using genetic algorithm. Phys. Med. Biol. **49**, 1915–1932 (2004)
18. Schreibmann, E., Lahanas, M., Xing, L., Baltas, D.: Multiobjective evolutionary optimization of the number of beams, their orientations and weights for intensity-modulated radiation therapy. Phys. Med. Biol. **49**, 747–770 (2004)
19. Bortfeld, T., Schlegel, W.: Optimization of beam orientations in radiation therapy: some theoretical considerations. Phys. Med. Biol. **38**, 291–304 (1993)
20. Djajaputra, D., Wu, Q., Wu, Y., Mohan, R.: Algorithm and performance of a clinical IMRT beam-angle optimization system. Phys. Med. Biol. **48**, 3191–3212 (2003)
21. Lu, H.M., Kooy, H.M., Leber, Z.H., Ledoux, R.J.: Optimized beam planning for linear accelerator-based stereotactic radiosurgery. Int. J. Radiat. Oncol. Biol. Phys. **39**, 1183–1189 (1997)
22. Pugachev, A., Xing, L.: Incorporating prior knowledge into beam orientation optimization in IMRT. Int. J. Radiat. Oncol. Biol. Phys. **54**, 1565–1574 (2002)
23. Rowbottom, C.G., Oldham, M., Webb, S.: Constrained customization of non-coplanar beam orientations in radiotherapy of brain tumours. Phys. Med. Biol. **44**, 383–399 (1999)
24. Stein, J., Mohan, R., Wang, X.H., Bortfeld, T., Wu, Q., Preiser, K., Ling, C.C., Schlegel, W.: Number and orientations of beams in intensity-modulated radiation treatments. Med. Phys. **24**, 149–160 (1997)
25. Chen, G.T., Spelbring, D.R., Pelizzari, C.A., Balter, J.M., Myrianthopoulos, L.C., Vijayakumar, S., Halpern, H.: The use of beam's eye view volumetrics in the selection of non-coplanar radiation portals. Int. J. Radiat. Oncol. Biol. Phys. **23**, 153–163 (1992)
26. Cho, B.C.J., Roa, H.W., Robinson, D., Murray, B.: The development of target-eye-view maps for selection of coplanar or noncoplanar beams in conformal radiotherapy treatment planning. Med. Phys. **26**, 2367–2372 (1999)
27. Goitein, M., Abrams, M., Rowell, D., Pollari, H., Wiles, J.: Multi-dimensional treatment planning: II beam's eye-view, back projection, and projection through CT sections. Int. J. Radiat. Oncol. Biol. Phys. **9**, 789–797 (1983)
28. Pugachev, A., Xing, L.: Computer-assisted selection of coplanar beam orientations in intensity- modulated radiation therapy. Phys. Med. Biol. **46**, 2467–2476 (2001)
29. Pugachev, A., Xing, L.: Pseudo beam's-eye-view as applied to beam orientation selection in intensity- modulated radiation therapy. Int. J. Radiat. Oncol. Biol. Phys. **51**, 1361–1370 (2001)
30. Das, S.K., Marks, L.B.: Selection of coplanar or noncoplanar beams using three-dimensional optimization based on maximum beam separation and minimized nontarget irradiation. Int. J. Radiat. Oncol. Biol. Phys. **38**, 643–655 (1997)
31. D'Souza, W.D., Zhang, H.H., Nazareth, D.P., Shi, L., Meyer, R.R.: A nested partitions framework for beam angle optimization in intensity-modulated radiation therapy. Phys. Med. Biol. **53**, 3293–3307 (2008)
32. Zhang, H.H., Shi, L., Meyer, R.R., Nazareth, D., D'Souza, W.: Solving beam-angle selection and dose optimization simultaneously via high-throughput computing. INFORMS J. Comput. **21**(3), 427–444 (2009)
33. Lee, E.K., Fox, T., Crocker, I.: Simultaneous beam geometry and intensity map optimization in intensity-modulated radiation therapy treatment planning. Ann. Oper. Res. **119**, 165–181 (2003)
34. Lee, E.K., Fox, T., Crocker, I.: Integer programming applied to intensity-modulated radiation therapy treatment planning. Int. J. Radiat. Oncol. Biol. Phys. **64**, 301–320 (2006)
35. Gokhale, P., Hussein, E.M., Kulkarni, N.: The use of beam's eye view volumetrics in the selection of non-coplanar radiation portals. Med. Phys. **23**, 153–163 (1994)
36. Meedt, G., Alber, M., Nüsslin, F.: Non-coplanar beam direction optimization for intensity-modulated radiotherapy. Phys. Med. Biol. **48**, 2999–3019 (2003)
37. Wang, X., Zhang, X., Dong, L., Liu, H., Gillin, M., Ahamad, A., Ang, K., Mohan, R.: Effectiveness of noncoplanar IMRT planning using a parallelized multiresolution beam angle optimization method for paranasal sinus carcinoma. Int. J. Radiat. Oncol. Biol. Phys. **63**, 594–601 (2005)
38. Holder, A., Salter, B.: A tutorial on radiation oncology and optimization. In: Greenberg, H. (ed.) Tutorials on Emerging Methodologies and Applications in Operations Research. Kluwer Academic Press, Boston, MA (2004)

39. Li, Y., Yao, J., Yao, D., Chen, W.: A particle swarm optimization algorithm for beam angle selection in intensity-modulated radiotherapy planning. Phys. Med. Biol. **50**, 3491–3514 (2005)
40. Morrill, S.M., Lane, R.G., Jacobson, G., Rosen, I.I.: Treatment planning optimization using constrained simulated annealing. Phys. Med. Biol. **36**, 1341–1361 (1991)
41. Oldham, M., Khoo, V., Rowbottom, C.G., Bedford, J., Webb, S.: A case study comparing the relative benefit of optimising beam-weights, wedge-angles, beam orientations and tomotherapy in stereotactic radiotherapy of the brain. Phys. Med. Biol. **43**, 2123–2146 (1998)
42. Rowbottom, C.G., Webb, S., Oldham, M.: Improvements in prostate radiotherapy from the customization of beam directions. Med. Phys. **25**, 1171–1179 (1998)
43. Rowbottom, C.G., Webb, S., Oldham, M.: Beam-orientation customization using an artificial neural network. Phys. Med. Biol. **44**, 2251–2262 (1999)
44. Söderstrom, S., Brahme, A.: Selection of suitable beam orientations in radiation therapy using entropy and fourier transform measures. Phys. Med. Biol. **37**, 911–924 (1992)
45. Wang, X., Zhang, X., Dong, L., Lie, H., Wu, Q., Mohan, R.: Development of methods for beam angle optimization for IMRT using an accelerated exhaustive search strategy. Int. J. Radiat. Oncol. Biol. Phys. **60**, 1325–1337 (2004)
46. Woudstra, E., Heijman, B.J.M.: Automated beam angle and weight selection in radiotherapy treatment planning applied to pancreas tumors. Int. J. Radiat. Oncol. Biol. Phys. **56**, 878–888 (2004)
47. Aleman, D.M., Glaser, D., Romeijn, H.E., Dempsey, J.F.: A primal-dual interior point algorithm for fluence map optimization in intensity modulated radiation therapy treatment planning. Phys. Med. Biol. **55**(18), 5467–5482 (2010)
48. Bertsekas, D.P.: On the Goldstein-Levitin-Polyak gradient projection method. IEEE T Automat. Contr. **21**(2), 174–184 (1976)
49. Nocedal, J., Wright, S.J.: Numerical Optimization. Springer Science, (2006).
50. Jitprapaikulsarn, S.: An optimization-based treatment planner for Gamma Knife radiosurgery. PhD thesis, Case Western Reserve University, May (2005).
51. Leksell, L.: Brain fragment. In: Steiner L. et al. and Raven Press Ltd (eds.) Radiosurgery baseline and trends (1992).
52. Leksell, L.: The stereotactic method and radiosurgery of the brain. Acta Chirurgica Scandinavica **102**, 316–319 (1951)
53. Ferris, M.C., Lim, J., Shepard, D.M.: Radiosurgery treatment planning via nonlinear programming. Ann. Oper. Res. **119**, 2002 (2001)
54. Ferris, M.C., Lim, J., Shepard, D.M.: An optimization approach for radiosurgery treatment planning. SIAM J. Optimiz. **13**, 921–937 (2002)
55. Cundy, H., Rollett, A.: Mathematical Models, 3rd edn. Tarquin Pub., Stradbroke, England (1989).
56. Rogers, C.A.: Packing and Covering. Cambridge University Press, Cambridge, England (1964)
57. Ferris, M.C., Lim, J., Shepard, D.M.: Optimization of Gamma Knife radiosurgery. In: Du D-Z., Pardalos P., Wang J. (eds.) Discrete Mathematical Problems with Medical Applications, vol. 55 of DIMACS Series in Discrete Mathematics and Theoretical Computer Science, pp. 27–44. American Mathematical Society (2000).
58. Luan, S.H., Swanson, N., Chen, Z.H., Ma, L.: Dynamic Gamma Knife radiosurgery. Phys. Med. Biol. **54**, 1579–1591 (2009)
59. Wu, Q.J., Chankong, V., Jitprapaikulsarn, S., Wessel, B.W., Einstein, D.B., Kinsella, T.J., Mathayomchan, B.: Real-time inverse planning for Gamma Knife radiosurgery. Med. Phys. **30**, (2004).
60. Wu, Q.J., Jitprapaikulsarn, S., Mathayomchan, B., Einstein, D.B., Maciunas, R.J., Pillai, K., Wessel, B.W., Kinsella, T.J., Chankong, V.: Clinical evaluation of a Gamma Knife inverse planning system. Radiosurg. **5**, (2004).
61. Ghobadi, K., Ghaffari, H.R., Aleman, D.M., Ruschin, M., Jaffray, D.A.: Automated treatment planning for a dedicated multi-source intra-cranial radiosurgery treatment unit using projected gradient and grassfire algorithms. Med. Phys. **39**(6), 3134–3141 (2012).
62. Oskoorouchi, M.R., Ghaffari, H.R., Terlaky, T., Aleman, D.M.: An interior point constraint generation algorithm for semi-infinite optimization with healthcare application. Oper. Res. **59**(5), 1184–1197 (2011)

Portable Asset Management in Hospitals

Nebil Buyurgan and Nabil Lehlou

1 Introduction

Portable asset management is a critical piece of a hospital's logistical operations. Large quantities of portable assets are being transported, utilized, operated and repositioned by a number of people to serve patients in environments that are highly sensitive to time. The asset management practice in hospitals essentially includes the localization and monitoring of resources through a maintained comprehensive list of locations and statuses of medical equipment (e.g. heart monitors, infusion pumps, wheelchairs, and air purification equipment). In this context, two major issues are identified in hospital environments related to the asset management: (1) equipment availability and utilization, and (2) the time required for their localization and acquisition. According to industrial and academic studies [1, 2], the average utilization rate for portable assets in hospitals ranges between 35% and 45%, depending on the management system models and the technology being used. In addition, according to a study performed by Dr. John Halamka, CIO of CareGroup and Harvard Business School, doctors and nurses spend an average of 20 min per day looking for misplaced medical equipment, at a cost of up to $100 an hour [3]. To improve the visibility of portable assets and decrease their underutilization rates, hospitals can enhance their management practices through real-time locating systems (RTLS). Different technologies are used to support RTLSs including Automatic Identification (AutoID) technologies such as barcode technologies, passive and active Radio Frequency Identification (RFID) systems, Geographic Information Systems (GIS) and Web-based GIS, short message services via wireless communication, and infrared (IR) technology. Industrial applications, whose functional goal is to enable automated portable equipment management, typically consist of

---

N. Buyurgan (✉) • N. Lehlou
BELL 4207, University of Arkansas,
Fayetteville, AR 72701, USA
e-mail: nebilb@uark.edu; nlehlou@uark.edu

networked entities that relay the necessary information to a main server. Authorized clinical and technical staff can then access this information using software. In this chapter, we present an analytical study of portable asset management practices in hospital environments using a simulation-based decision support tool. The tool takes into consideration different material handling and utilization processes and is used to investigate the impact of RFID technology on widely adopted organizational portable asset management models.

2 Common Portable Asset Management Models

Based on extensive literature reviews and numerous hospital system investigations, portable asset management systems in hospitals can typically be modeled as three main configurations: centralized, semi-centralized, and decentralized. The centralized and decentralized configurations are usually extreme cases for specific portable assets where the management has strict control or no control over medical equipment, respectively. For example, a centralized control over a certain type of medical equipment could be forced, if a hospital owns a few of that equipment. Another main driver for centralized control is equipment abuse by nurses and patients. Defibrillators and IV pumps are good examples of these types of equipment. On the other hand, a decentralized control can be used over the portable equipment that are (1) commonly used and (2) do not require sanitation after every use. Electronic thermometers, otoscopes, and wheel chairs are good examples of this type.

The semi-centralized configuration is a mixed model where some level of central control is enforced by the management. This is a common practice in hospitals due to different requirements of internal departments or units. A typical emergency department for example requires a certain number of wheelchairs readily available all the time. When a patient is transferred from the emergency department to another department, the wheelchair also leaves with him/her. The nurse usually brings the wheelchair back after the patient is transferred. In the case of IV pumps however, the device stays with the patient and the nurse has to find a replacement. In general, the three aforementioned configurations are used simultaneously in a hospital environment.

Typically, hospitals have one or more equipment centers. The number of centers depends on the hospital size and structure, the number of internal units and their locations (e.g. onsite vs. offsite), and the number of portable medical equipment to be traced. Another driver for the equipment management system is the type of portable asset. As previously mentioned, some medical equipment may require a complete sanitation before the next use, whereas cleaning with wet wipes may be enough for others. Some medical equipment may require changing certain pieces after every use; no additional cleaning may be necessary. In order to simplify the analysis of different hospital environments and their associated portable asset management systems in our study, a hospital is viewed as a set of departments called *units*. Based on the structure of units and the types of traced portable equipment, asset management systems and their associated processes are modeled as the following.

Portable Asset Management in Hospitals

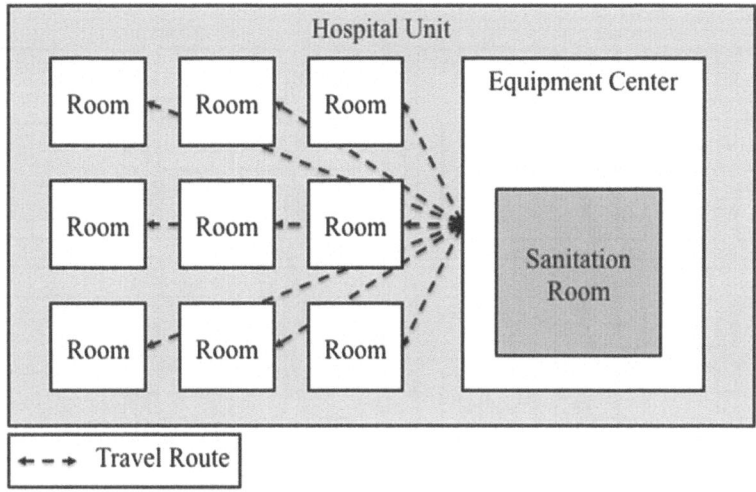

Fig. 1 Sample layout of a centralized system

2.1 Centralized Systems

Centralized portable asset management systems in hospitals typically have only one equipment center in every unit (a layout example of a unit is depicted in Fig. 1). In this type of systems, every unit can be considered as a self-sufficient and isolated entity from the other hospital units. In case of a need, a nurse checks out a cleanable piece from the unit's equipment center; he/she uses it, cleans it, and then returns it back to the center before the next checkout. The equipment center can be considered as a clean inventory room in this system in which the nurses perform all the equipment-related activities. Similarly, general procedure for the sanitizable equipment is also nurse-oriented in the centralized portable asset management systems. There is a sanitation room, typically located in every equipment center. Equipment is checked out from the unit's center, used, and then returned to the center by the nurse to be sanitized before the next checkout. They are sanitized by the sanitation personnel and stored in the equipment center.

2.2 Semi-Centralized Systems

Semi-centralized portable asset management systems in hospitals usually utilize hybrid models of both centralized and decentralized systems. This management model is the most commonly used in hospitals. A semi-centralized portable asset management system has a set of hospital units with a centralized management system in each. However, both cleanable and sanitizable equipment are shared among the units' equipment centers. Equipment is transferred from one unit to another when nurses do not find an available item in the unit to which he/she is assigned (his/her original unit). This creates a high-traffic material flow within the hospital

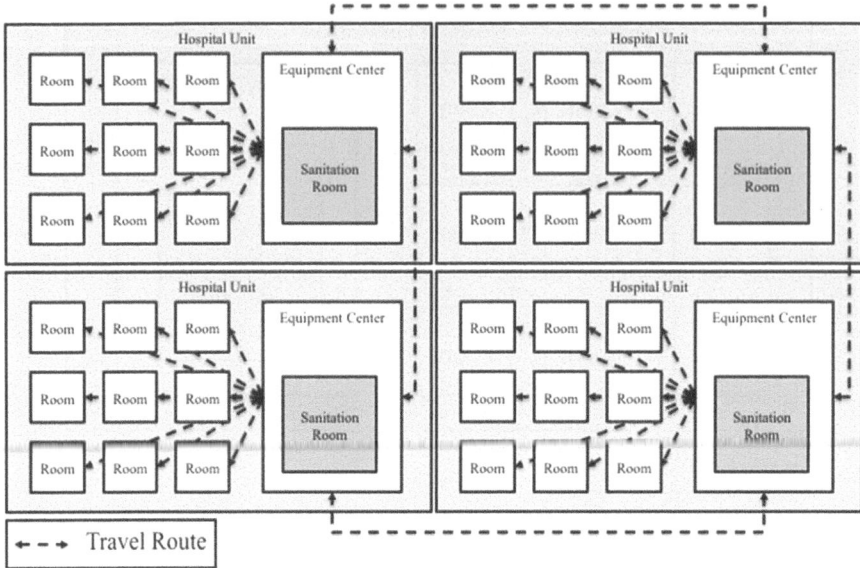

Fig. 2 Sample layout of a semi-centralized system

that leads nurses to search for equipment in different hospital units when none are available in the original unit. A layout example is depicted in Fig. 2. In this case, the nurse checks out an item from another unit, but returns it to the center of his/her unit. The cleaning and sanitation processes for the equipment are typically the same as the centralized portable asset management systems.

2.3 Decentralized Systems

Decentralized portable asset management systems in hospitals are on the extreme end of the equipment management and usually are used in very large size hospitals or multi-hospital systems. In a decentralized portable asset management system, there is one equipment center with one sanitation room from which sanitizable equipment is checked out, returned, sanitized, and stored; cleanable portable equipment is located in the hospital's rooms. In other words, while cleanable equipment is cleaned and left in the room where it was used, sanitizable equipment are picked by collectors from rooms and taken to the equipment center to be sanitized. These collectors perform their cycles periodically by visiting rooms, picking items that need sanitation, and taking them to the center to be sanitized. A layout example is depicted in Fig. 3.

These systems may or may not use the support of a technological infrastructure such as an RFID-based RTLS, which has a set of features that allows the organization to run portable asset management operations at a certain level of efficiency (see Table 1).

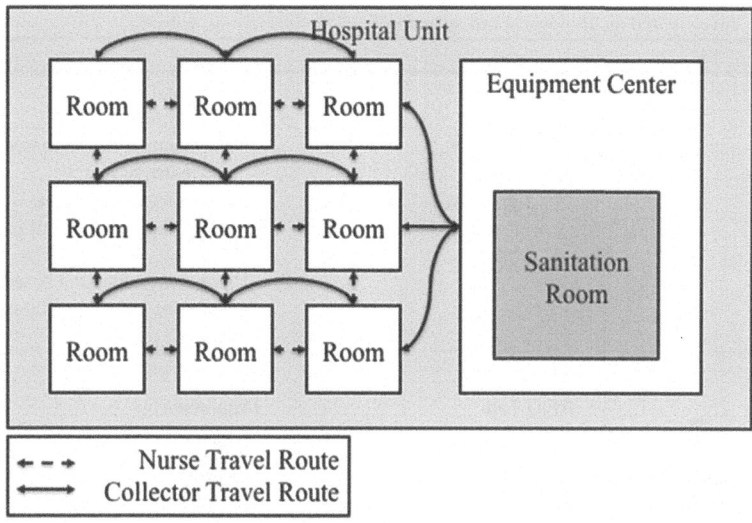

Fig. 3 Sample layout of a decentralized system

3 RFID and Portable Asset Management

Radio Frequency Identification is a data collection technology that utilizes wireless radio communication (radio frequency signals) to identify, track, and categorize objects (see Fig. 4). The basic RFID system consists of three main components:

- The RFID reader, which by itself contains the processing unit, antennas, and the cables joining them; its main task is to send electromagnetic waves to the surrounding environment and listen for electromagnetic responses from the RFID tags. Upon receipt of the tags' data, the reader submits the RFID reads to the target database.
- The RFID tag, which is a microchip that is bound to a small antenna and that transmits the data stored in it as the electromagnetic response to the reader.
- The database where all the raw read data is to be amassed, and maybe converted into meaningful numbers and patterns.
- This system can be extended with a set of middleware devices, a variety of soft-controllers, a network of readers, and a powerful database management system (DBMS) to ease data-acquisition and data-management in a large information system.

With its capability of storing a relatively large amount of data, an RFID tag can outperform a barcode tag, which can identify the kind of an item only, one item at a time, and has to be scanned with line of sight. When an RFID tag utilizes batteries to function, it is called *Active*, it can be read from far distances (up to 30 m, or 100 feet), and it uses a specific range of radio frequencies. *Passive* tags on the other hand

Table 1 Investigated technological infrastructures

Option	Cost	Features
Manual	Free	Traditional search for equipment
		Equipment collection based on exhaustive search
RFID	Expensive	Knowledge of the exact location and status of each item
		Equipment collection based on visiting necessary rooms only

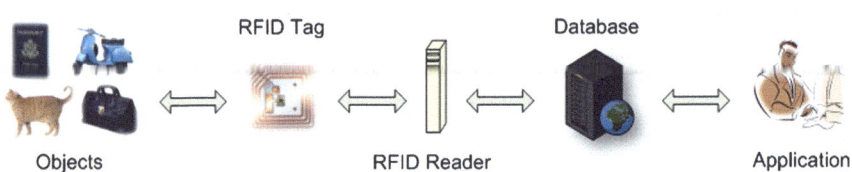

Fig. 4 Object/device interactions in an RFID system

do not require batteries; they are powered by the electromagnetic waves sent by the reader, and that is why their read-distance is limited to a number of feet. The main advantage that Passive technology has over its Active counterpart is the significant cost and maintenance reductions.

Along with its benefits and unique applications, RFID is one of the major enabling technologies that are considered and adopted by hospitals. Improving inventory management and asset control, increasing patient safety, reducing medical errors, and monitoring critical processes are some of the key drivers that motivate healthcare industry to invest in this technology. Recent studies report that the market for RFID tags and systems in healthcare will grow exponentially from $90 million in 2006 to $3.5 billion in 2017 [4]. The deployment of RFID technology in the healthcare industry can be classified into five major categories: portable asset management, inventory management, authenticity management, identity management and process management [5]. Highlighted by the industry experts, the most common application of RFID in healthcare is portable asset management.

According to a recent survey released by the Healthcare Information and Management Systems Society [6], almost a quarter of the healthcare industry believes that the widespread use of RFID applications benefit healthcare organizations in the area of asset and biomedical equipment safety. The same survey reveals that asset and biomedical equipment tracking is the top area in which RFID is being used in healthcare organizations. Specific portable asset management implementations in hospitals include real-time equipment count and tracking, inspection/maintenance scheduling and management, hygiene compliance management, and recall/return management. From business case development and cost-benefit analysis point of view, we will discuss a handful of applications.

Wayne Memorial Hospital, located in Goldsboro, North Carolina, USA, has deployed RFID technology in its real-time asset location and management program. With the objective to help the staff keep tabs on the location and status of tagged assets, the hospital tagged and tracked IV pumps (i.e., devices used to introduce medicines, liquids, etc. into a patient's body), diagnostics machines, blood warmers, wheelchairs, etc. Once the organization collected data from the system, they found that only 50-60% of their infusion pumps were being utilized. Subsequently, the hospital reduced the number of pumps purchased from 300 to 250, which saved them about $276,000 and as well as an additional about $27,000 in maintenance cost. In total, Wayne Memorial saved more than $303,000 by improving the visibility of infusion pumps through RFID technology. In addition, the hospital saved about $24,000 by tracking other equipment. A total of about $327,000 has been realized as capital expense reduction [7].

Brigham and Women's Hospital, located in Boston, Massachusetts, USA, has implemented a hospital-wide real-time location platform to manage 9,854 medical devices, including 2,500 across central transport. Room- and zone-level coverage has been utilized throughout care areas on 17 hospital floors including perioperative and emergency departments. The hospital projected $300,000 yearly gross savings that reduced return on investment down to 1 year. In addition, with the use of the technology, the hospital realized increased staff satisfaction and productivity, increased efficiency, reduced loss; and improved equipment flow based on real-time alerts and finding equipment quickly [8]. A potential side benefit from that implementation was a significant reduction in the length of patient stay. The hospital realized an average of 0.1 day shorter patient stay by reducing portable asset locating time, which triggered 700 more discharges and $10 million in revenue.

PinnacleHealth Hospital system, located in Pennsylvania, USA, has utilized an RFID-based real-time location system for its hospital-wide asset tracking activities. The hospital system has deployed RFID tags to locate as many as 10,000 devices, which made this among the largest RFID deployments in healthcare. They have used both wired and wireless receivers as well as reusable RFID tags for small equipment, rental equipment and patient tracking to reduce installation costs. Hospital staff can track equipment across and between the two sites. The return on investment from real-time asset tracking came in 12 months for PinnacleHealth's flagship 546-bed Harrisburg Hospital, representing $900,000 in savings [9].

A list of US healthcare providers that have reported on the implementation of a hospital-wide RFID system is summarized below.

- University Medical Center in Tucson Arizona has installed one of the largest WiFi-based asset management systems covering eight floors, over a million square feet and 2,300 tagged assets that included a wide range of portable devices. Hospital staff members have used a turnkey solution to track and manage equipment such as infusion pumps, beds, monitors, wheelchairs and other portable devices [10].
- The Bon Secours Health System, in Richmond, Virginia, has installed RFID at three hospitals to track 12,000 pieces of equipment. The hospital has reported

that the nursing staff have saved 30 min per shift, on average, in searching for equipment and they have dramatically reduced the number of missing portable devices [11].
- Advocate Good Shepherd Hospital, Barrington, Illinois, Holy Name Society Hospital, Teaneck, New Jersey, St. Vincent's Hospital, Birmingham, Alabama have used RFID technology to track portable assets and surgical instruments as they move around the hospital. The data on their current location, last sterilization time, and maintenance records has been collected [1].

4 Portable Asset Management Systems Analysis

In this section, an analysis of portable equipment management in hospital environments is presented. Considering the significant amount of personnel time for equipment search, described portable asset management models, tracking activities, and technology involvement in those activities are investigated. A simulation-based decision support tool is constructed to analyze the different processes and the impact of RFID technology on the widely adopted organizational systems. In order to conduct the analysis, a Java simulation-based model is developed in a flexible way to allow different variable entries. System variables include the type of the portable asset management system model, its technological framework, the number of rooms and units, the number of portable equipment being tracked, the demand associated with them, the expected utilization time, and the time needed to travel from one location to another.

Since the main difference between the systems is the way operations are performed, the built simulation model is a task-oriented one. To be more specific, a hospital is formed from one or more units (e.g. emergency department, intensive care unit, OR department, etc.), where each unit, equipment center, and room is associated with a list of tasks to be performed by the nurses or the collectors. Also, since the personnel are the ones seizing the portable equipment resources, they are generated according to the given demand rate in a certain unit of a hospital. That leads them to follow their corresponding task-list before leaving the simulated system. The constructed tasks that are used in the simulation model are given in Table 2.

In the case of a centralized portable asset management system, the technology being used does not make a difference in processes because the associated process for each nurse remains the same regardless of the underlying technological framework. Table 3 shows such a process.

The processes for nurses in a semi-centralized portable asset management system with no infrastructure and with RFID technology are described in Table 4. Note that when RFID technology is integrated in this system, personnel can look up pieces of equipment and their associated location and status before traveling to acquire them.

The processes for nurses and collectors in a decentralized portable asset management systems with no infrastructure are described in Table 5. When RFID

Table 2 Constructed tasks used in the simulation model

Task	Description
Delay	Helps advance the simulation clock due to the performance of an operation
Seize	An event that allocates a resource to a nurse, and transfers it from an external unit to the nurse's original unit if these units are different
Release	An event that frees an allocated resource and allocates it to the next nurse waiting in line (if there is one)
TrySeize	A branching mechanism that performs a Seize if possible. If the Seize is successful, the next task in the sequence is performed; else, an alternative task is carried out (usually a search)
Travel	Directs an entity (nurse or collector) to a destination, and then schedules a delay to model the trip duration
Reoriginate	A Travel to the original place (unit/center/room) of an entity
Search	A combination of a Travel and a TrySeize that helps visiting new places until a tool is obtained
Find	A set of lookups and Travels: If a nurse finds an available resource when performing a lookup, s/he travels to it to acquire it. If s/he does not find it when s/he arrives to the destination, s/he redoes the lookup. If s/he does not find any available resource at all, s/he goes and waits in line in the original unit
Iterate	A loop mechanism that helps iterate over a list of tasks.
Accumulate	Helps leaving equipment in a certain place (the original place of an entity) without releasing it
Pickup	An event that releases all allocated resources (by entities: nurses) at a certain place and allocates them to another entity (the collector)
Collect	A set of travels to and pickups from all the places of a hospital
LookupCollect	A set of travels to and pickups from places where unsanitized equipment is located
ReleaseAll	An event that releases a whole set of resources at once, such as the case of getting equipment out of a sanitizing room

technology is integrated, two things change: the Search task becomes a Find task and the Collect task becomes a LookupCollect task; and that is because RFID allows personnel to know where equipment is before they start a trip to go seize it.

To compare the efficiency of the three systems and to investigate the impact of the integration of RFID technology in hospitals, the simulation models discussed above are used on the same hospital. However, it is hard to find a large-scale real-world hospital, where all systems can be adopted, as well as collect the time required to move from one place to another. In the short-term of this research, it is opted to simplify the task of acquiring a space-model by constructing a synthetic hospital.

Table 3 Processes in a centralized system

Centralized			
No technology or RFID technology			
Cleanable		Sanitizable	
Task	Significance	Task	Significance
Delay	Go to center	Delay	Go to center
Seize	Check out item	Seize	Check out item
Delay	Go to room	Delay	Go to room
Delay	Use item	Delay	Use item
Delay	Clean item	Delay	Go to center to return item
Delay	Go to center	Delay	Sanitize item
Release	Return item	Release	Make item available

This hospital consists of 10 units where each one has 36 rooms and one equipment center, as depicted by Fig. 5.

Note that in the decentralized system case, only one center is used for sanitation whereas the other nine are not visited at all. From the temporal aspect of trips, it takes one time period to move from one room to its adjacent places, whether the move is horizontal or vertical (a time period of 15 s is used.) Diagonal moves are not allowed; therefore, to move from Room 1 to Room 8 for example, it takes two time periods since the individual has to perform one horizontal step and one vertical step. Even though our used software library does not support spatial representation just yet, this hospital layout helps us generate the matrix of travel-durations from one place to another.

In the experimentation phase, each simulation model is run for 10 replications with a warm-up period of 50 simulation minutes and a runtime of 10,080 simulation minutes (1 week) with the input data presented in Table 6 (time values are in simulation minutes.) The performance measure is the time spent to acquire an item in the hospital (time-to-acquire) by a nurse or by a collector. The simulation results are shown in Table 7 and graphed in Fig. 6.

According to these results, the decentralized portable asset management system seems to be the most efficient one overall for cleanable and sanitizable equipment. This is true whether there is a technological infrastructure in the system or not. The efficiency gain for nurse travel time within the decentralized system is greater than the collector travel time due to the effect of RFID technology. Between cleanable and sanitizable cases, the efficiency gains for nurse travel times are not statistically significant. Due to the simulated processes, there is no efficiency gain in the centralized portable asset management systems for incorporating technological infrastructure into the system.

The decentralized portable asset management system simulation results are significant. The semi-centralized system with no infrastructure is the least efficient system overall for cleanable and sanitizable equipment. The impact of RFID technology in the system is more significant in this system than any other one due to the significant reduction in the performance measure (about 38%). For the collector's

Table 4 Processes in a semi-centralized system

Semi-centralized				RFID technology			
No technology							
Cleanable		Sanitizable		Cleanable		Sanitizable	
Task	Significance	Task	Significance	Task	Significance	Task	Significance
Delay	Go to center	Delay	Go to center	Find	Look up available item and go acquire it	Find	Look up available item and go acquire it
TrySeize	Try to obtain item from the center	TrySeize	Try to obtain item from the center	Reoriginate	Return to original unit	Reoriginate	Return to original unit
Search	Search for item in case no item is found	Search	Search for item in case no item is found	Delay	Go to original room	Delay	Go to original room
Reoriginate	Return to original unit	Reoriginate	Return to original unit	Delay	Use item	Delay	Use item
Delay	Go to original room	Delay	Go to original room	Delay	Clean item	Delay	Go to center to return item
Delay	Use item	Delay	Use item	Delay	Go to center	Delay	Sanitize item
Delay	Clean item	Delay	Go to center to return item	Release	Return item	Release	Make item available
Delay	Go to center	Delay	Sanitize item				
Release	Return item	Release	Make item available				

Table 5 Processes in a decentralized system

Decentralized					
No technology					
Nurses generated in a room				Collector generated in a center	
Cleanable		Sanitizable		Sanitizable	
Task	Significance	Task	Significance	Task	Significance
TrySeize	Try to obtain item from original room	Travel	Go to center	Delay	Wait before next round
Search	Search for item in case no item is found	Seize	Check out item	Collect	Pick up used items from rooms
Reoriginate	Travel to original room	Reoriginate	Go to original room	Reoriginate	Go to center
Delay	Use item	Delay	Use item	Delay	Sanitize items
Delay	Clean item	Accumulate	Leave tool in the room	ReleaseAll	Make items available
Release	Leave item in the room			Iterate	Repeat from 1st task

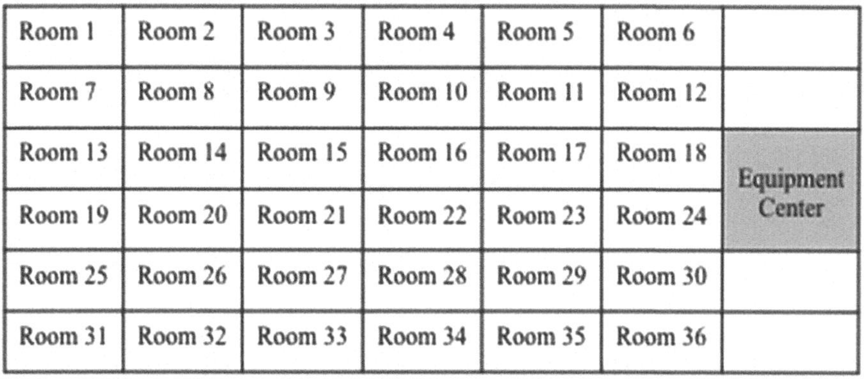

Room 1	Room 2	Room 3	Room 4	Room 5	Room 6	
Room 7	Room 8	Room 9	Room 10	Room 11	Room 12	
Room 13	Room 14	Room 15	Room 16	Room 17	Room 18	Equipment Center
Room 19	Room 20	Room 21	Room 22	Room 23	Room 24	
Room 25	Room 26	Room 27	Room 28	Room 29	Room 30	
Room 31	Room 32	Room 33	Room 34	Room 35	Room 36	

Fig. 5 Layout of a synthetic hospital

Table 6 Simulation input data

Variable	Value/distribution
Number of units	10
Number of rooms	36
Number of items	40
Demand rate	1 item per 5 min
Usage time	Triangular (10,15,20)
Sanitization	Triangular (5,7,10)
Cleaning	Triangular (2,3,4)

time to acquire an item, there is an improvement with the introduction of RFID, but it is not statistically significant.

In conclusion and according to the studied simplified model, if a hospital cannot afford a technological infrastructure, it is recommended to implement a decentralized system. If a hospital has a semi-centralized system, the hospital is urged to integrate RFID technology in its system to move from the least efficient system to the third best. Finally, if a hospital is under construction, it is advised to avoid the implementation of a centralized system for it is not efficient while there is no room for improvement.

5 Summary

This chapter focuses on evaluating different hospital systems (centralized, semi-centralized, and decentralized) that can either have no technological infrastructure or have an RFID-based RTLS, which enables tracking equipment and its status in hospitals. The systems are studied thoroughly and the process through which the associated equipment goes through is identified. A task-oriented approach is used to model these processes in a Java simulation model. According to obtained results,

Table 7 Results of simulating the three hospital systems

System	Infrastructure	Equipment	Avg. Time-to-acquire (min)	St. dev.
Centralized	None	Cleanable	15.54	0.2368
		Sanitizable	15.58	0.3573
	RFID	Cleanable	15.54	0.2368
		Sanitizable	15.58	0.3573
Semi-centralized	None	Cleanable	16.38	0.5594
		Sanitizable	16.39	0.8945
	RFID	Cleanable	10.25	0.8541
		Sanitizable	10.06	0.5874
Decentralized	None	Cleanable	8.06	1.6064
		Sanitizable–nurse	11.20	0.4315
		Sanitizable–collector	7.56	0.7262
	RFID	Cleanable	5.33	1.6184
		Sanitizable–nurse	8.31	0.2858
		Sanitizable–collector	7.38	0.1581

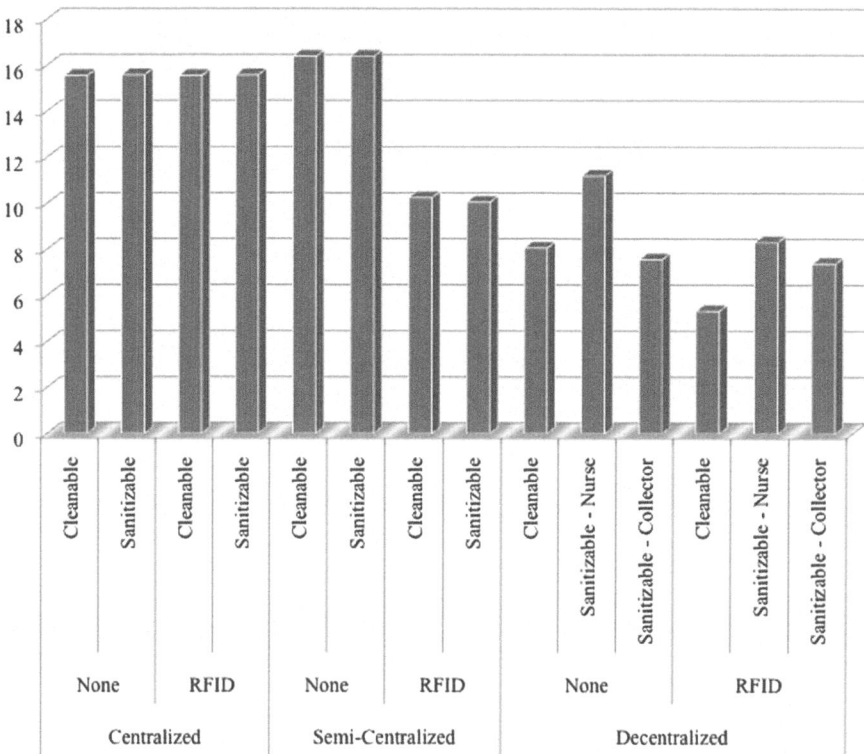

Fig. 6 Results of simulating the three hospital systems

the best system in terms of nurse's average time to reach the equipment is the decentralized system with an RFID framework. This is true for both cleanable and sanitizable equipment. Also, the integration of RFID technology does make a difference in both the decentralized and semi-centralized systems, with the semi-centralized one benefiting the most from the technology.

References

1. Vilamovska, A.M., Hatziandreu, E., Schindler, H.R., van Oranje-Nassau, C., de Vries, H., Kraples, J.: Study on the requirements and options for RFID application in healthcare-identifying areas for radio frequency identification deployment in healthcare delivery: a review of relevant literature, Report, RAND Corporation Europe (2009).
2. Hill-Rom.: Hill-Rom's AssetAdvantage® Equipment Utilization and Nursing Satisfaction Site Assessment. http://www.hill-rom.com/usa/PDF/155517.pdf (2008) Accessed 1 Sept 2010.
3. Halamka, J.: RFID: Lessons learned about innovation, infrastructure, and ROI. EMC Business Views. http://www.emc.com/leadership/business-view/rfid.htm (2008) Accessed 26 Aug 2010.
4. Harrop, P.: Rapid adoption of RFID in healthcare. IDTechEx. http://www.idtechex.com/products/en/articles/00000470.asp (2006) Accessed 25 June 2010.

5. Buyurgan, N., Hardgrave, B.C., Lo, J., Walker, R.T.: RFID in healthcare: a framework for uses and opportunities. Int. J. Adv. Pervasive Ubiquitous Comput. **1**(1), 1–25 (2009).
6. Healthcare Information and Management Systems Society Vantage Point.: Use of RFID technology. http://www.himss.org/content/files/vantagepoint/vantagepoint_201006.asp?pg=1 (2010) Accessed 1 July 2010.
7. RAND Corporation.: Study on the requirements and options for radio frequency identification (RFID) application in healthcare. Technical Report. http://www.rand.org/pubs/technical_reports/2009/RAND_TR608.1.pdf (2009) Accessed 22 June 2010.
8. Bacheldor, B.: Brigham and Women's Hospital becomes totally RTLS-enabled. RFID J. http://www.rfidjournal.com/article/view/3931 (2008) Accessed 12 Aug 2010.
9. Radianse.: PinnacleHealth. http://www.radianse.com/success-stories-pinnacle.html (2010) Radiance Corp. Accessed 9 Aug 2010.
10. Aeroscout.: Philips announces new RFID asset tracking solution. http://www.aeroscout.com/content/philips-announces-new-rfid-asset-tracking-solution/philips-announces-new-rfid-asset-tracking (2010). Accessed 28 Aug 2010.
11. Swedberg, C.: Bon Secours Richmond finds RFID saves $2 million annually. RFID J. http://www.rfidjournal.com/article/view/7259 (2009) Accessed 28 Aug 2010.

Stochastic Integer Programming in Healthcare Delivery

Camilo Mancilla and Robert H. Storer

1 Introduction

We have identified two main research areas in healthcare delivery where stochastic integer programming has had an impact, resource allocation and operation management. Healthcare resource allocation is the study of problems related to finding an efficient allocation of healthcare resources, and is often referred to as strategic decisions in the operation research literature. In these problems the goal is typically to find designs that optimize services levels such as maximum coverage area or minimum travel time and at the same time minimize cost. Examples of these kinds of studies include ambulance allocation and healthcare facility location. The second general research area is healthcare operations management. This research concentrates on finding more efficient resource use within healthcare facilities. Scheduling operating rooms and assigning nurses to patients are the two main applications of stochastic integer programming in the operations management area.

Given the great diversity of patients and their medical needs, it is imperative to explicitly consider uncertainty in modeling and planning in healthcare delivery systems. Easily identifiable stochastic components include service times (duration of surgeries, exams, and other procedures), demand for service (arrival of emergency cases, no-show patients), and availability of resources (absentee staff, beds not vacated as scheduled, equipment malfunctions). Of direct interest in this paper are problems where decisions include an integer component such as the sequencing of surgeries, opening new facilities, or assigning nurses to patients. The aim of this manuscript is to review different applications of stochastic integer programming to facilitate the decision making process in healthcare delivery.

C. Mancilla (✉) • R.H. Storer
Lehigh University, Bethlehem, PA, USA
e-mail: cam306@lehigh.edu

2 Stochastic Integer Programming

Stochastic programming is a branch of optimization where some problem parameters are assumed uncertain. One of the assumptions in the stochastic programming literature is that we know the underlying joint distribution D of the set of uncertain parameters. Often a sample average approach is taken in which a sample of outcomes (or scenarios) is generated, and optimization of the average objective function is performed (with respect to the scenarios). Assuming the joint distribution D is known allows us to generate a representative set of scenarios. In some cases, sample data may exist (e.g., data may exist on the duration of surgical procedures in an OR scheduling problem). In this case, scenarios could be generated directly from data, thus eliminating the need to explicitly assume a distribution.

Stochastic programming problems can be classified by the number of decision points in time (or stages) where the decision maker is allowed to make a decision. If the decisions are made twice, with random variable outcomes occurring in between, we have a two-stage problem. If multiple decisions are made over time, with new random variable outcomes occurring in between, the problem is called a multistage problem.

For the two-stage stochastic programming problems, the decision variables are partitioned into two sets (first stage and second stage). The first stage variables are those that are decided before the outcome of the uncertain parameters becomes known. The second stage decision variables, or recourse actions, are those that are decided after observing the outcome of the random variables. In order to evaluate the selection of these first stage decision variables an objective function is defined. Often, neutral behavior with respect to the risk is assumed, therefore the natural objective function is expected cost.

$$\min_{x \in X}\{f(x) = E[F(x, D)]\} \quad (1)$$

As discussed in [22] it makes sense to utilize expected value in the objective function if the decision process is to be repeated many times as the law of the large numbers guarantees that, for a given (fixed) decision x, the average of the total cost, over many repetitions, will converge (with probability one) to the expectation $E[F(x, D)]$. Having optimized the selection of x in (1) under the feasible region X, we will have the optimal solution on average (in the long run) for our process.

In cases where the decision process requires binary or integer decisions in the first or second stages, the problem is said to be a stochastic integer programming problem.

In some problems the decision maker seeks a solution that does not exceed certain threshold τ with a specified probability:

$$P(F(x,D) > \tau) \geq \alpha \quad (2)$$

Constraints of this type are known as chance constraints. Example could include constraints that assure the probability of having an ambulance available in every period of some planning horizon or the probability of having an operation room available for emergency surgeries at any moment in the day always exceed a given service level threshold. A problem is called a stochastic integer problem with chance constraints when the computation of the probability involves the use of integer variables.

A multistage stochastic programming problem occurs when the decision maker makes decisions in a dynamic fashion over time. He/she will observe the system at multiple times t and, based on the information available at time t, will make a new decision. Detailed formalizations of these three types of problems can be found in [14, 22] sense to utilize. If some or all of the decision variables are restricted to integer values, the problem is said to be a stochastic multistage integer problem.

3 Healthcare Resource Allocation Problems

Locating facilities in health care often differ from other facility location problems that have been well studied in the location literature. For example, as opposed to minimizing logistics costs in standard problems, healthcare location problems may seek to guarantee service levels or maximize benefit to a community. Reference [5] provides a rich set of examples of the application of facility location in health care. The application of stochastic programming technique in healthcare facility location includes the allocation of ambulances and medical facilities.

3.1 Allocating Ambulances Using Stochastic Integer Programming

The problem consists of a set of ambulances and a set of potential customers (emergency patients) located in a two-dimensional space. The decisions involve how to allocate these ambulances in space such that a service level objective is maximized. The service level is typically measured as coverage of demand. This coverage of demand can be defined as the number of patient that can be reached within a given period of time. To illustrate the problem we will use Fig. 1 (this figure was provided in [9]). It presents demand nodes and potential ambulance bases (nodes indexed by numbers). Undoubtedly, the demand nodes are uncertain because nodes represent a demand aggregation of finite number of emergency request, for instance a city might be a node that represent thousand of emergencies requests in a given period of time.. The travel times are also assumed uncertain. Due to the incorporation of randomness directly into the models, this problem has an extra level of complexity relative to traditional facility location models. An excellent literature review of this type of problems is provided by [3].

In [12] the authors considered the problem of allocating ambulances to different stations in order to cover a given area. They assumed that the travel times from stations to each demand node follow a known distribution. They also model the fact that there might be delays between receiving a call and ambulance dispatching.

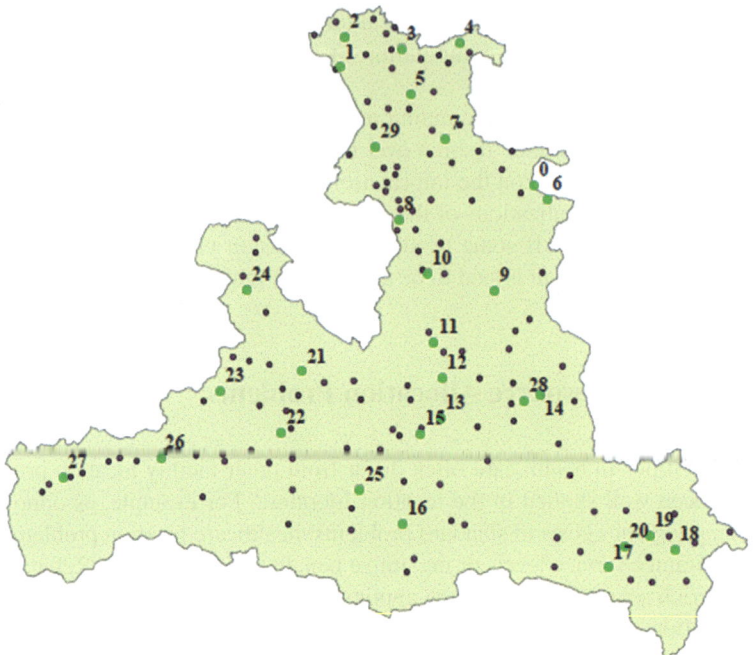

Fig. 1 Map of the providence of Salzburg (source [9])

Their model also allows not all calls to be served in cases where they do not have ambulance availability. They modeled the problem using a two-stage stochastic integer program where the recourse function is non-linear. They proved that the recourse function is concave in the set defined by the number of ambulances allocated by station when the number of ambulances is fixed. Since the objective function is concave they can conclude that the linear relaxation of their problem is a convex problem therefore a local minimum is a global minimum. They proposed a heuristic approach for the case where the number of ambulances is not fixed using the concavity results. The authors provided a computational study based on data provided by the towns of St. Albert and the City of Edmonton, Canada. They found that incorporating the randomness of pre-travel delays has an important impact on the solution in their model.

In [11] the combined problem of scheduling ambulance crews taking into consideration the maximum coverage throughout a planning horizon was studied. For the subproblem of allocating ambulances to maximize the expected coverage with probabilistic response times, the authors developed a tabu search heuristic algorithm. They also created two integer programs that use the output of the tabu search in order to find good solutions for the main problem. The first integer program tried to maximize the aggregate expected coverage while the second model was a biobjective model in which the first objective maximized the minimum expected coverage over every hour while the second objective was the same as that of the first model.

They tested their approach using real data from a Canadian EMS operator in a mid-size Canadian city. They found that the second integer program outperformed the first because the second integer program manages the concept of equity for every hour with a marginal deviation from maximum aggregate performance and the processing times of both integer programs were comparable.

3.2 Medical Facility Location

Since 1960, operation researchers have studied facility location problems, thus there is a rich literature in this regard. Reference [5] is a good coverage of facility location problems in health care. In medical facility location problems the decision maker is usually interested in maximizing patient coverage, and also reducing patient travel time to the facility. In [20] the problem of designing a network of senior centers is studied. The problem consists of deciding the locations of both comprehensive and satellites centers assuming that the demand is random for different sites. The state government would like to minimize the travel distance of the senior citizens and also the cost of opening facilities. The authors created a two-stage stochastic integer programming model with mixed-integer recourse. The solution approach was based on decomposition. First, they decomposed the problem by service regions and then they used scenario decomposition to solve the service region subproblems. In scenario decomposition non-anticipitivity constraints are relaxed using Lagrange relaxation. In order to close the duality gap they utilized a Lagrangian dual-based branch-and-bound algorithm proposed by [4] that successively reestablishes the equality of the first-stage vector . They also proposed a heuristic that reestablishes feasibility in the solution found by the decomposition by service region. They tested their solution with data from the census and GIS. They computed the expected value of perfect information in order to justify the use of stochastic integer programming and found the value to be around $200,000 per year.

4 Healthcare Operations Management

Hospitals must manage many capital intensive resources while guaranteeing quality service for their patients and satisfying their surgeons and doctors. In most US hospitals surgeons are not employees of the hospital but rather are independent service providers. The main reason surgery patients choose a particular hospital is that it is the hospital with which their surgeon is affiliated. Figure 2 illustrates the patient flow in a typical hospital. Patients typically arrive to the hospital for one of the following reasons:

A. Emergency
B. Consultation
C. Elective surgery
D. Exams (MRI, SCAN, blood test)

Emergency patients might go to the emergency room or directly to an operating room. Elective surgery patients would normally go to an operating room and then to a recovery room (PACU or ICU) after the procedure is performed. In both cases the hospital administration has to coordinate the downstream resources such as beds, staffing, etc. Patients coming for an exam or consult will typically go home afterwards. Coordinating and managing all these activities require a great deal of effort. One of the main challenges is the presence of uncertainty in patient demand, patient care, surgery duration, exam duration, consultation duration, case cancellation, recovery time in PACU or ICU, etc. Further, hospital management essentially has two customers, patients and physicians. This last consideration makes the use of mathematical models more challenging since one must consider multiple objectives (e.g., from the perspective of both patients and physicians). For example many decisions impact patient waiting time, physician idle time, and resource utilization simultaneously.

4.1 Operating Room Optimization

One of the main revenues drivers in a hospital are the operating rooms (ORs). According to [1] the ORs account for about 40% of the total revenue in a hospital, and operate at only 69% of their capacity. Therefore much effort has been devoted to investigating optimization methods that help improve their efficiency. According to [10] most hospitals follow one of two predominant ways to allocate time in the operation room: open block scheduling and block-booking scheduling. In open block scheduling surgeons submit requests for OR time on a daily basis and then the hospital schedules the surgeries based on capacity. In block-booking each surgeon or surgical team is allocated a block of uninterrupted OR time on a periodic basis (e.g., 7 a.m. to 3 p.m. every Tuesday and Thursday). In block-booking the surgeons (or their office staff) schedule patients into their own blocks subject to hospital approval.

4.2 Operating Room Planning

In the block-booking scheme hospital management must decide how much time to allocate to each surgeon or surgical team over a horizon (often 3 months). In [23] the authors used stochastic integer programming (SIP) to create a cyclic master schedule that will specify for each OR–day combination the amount of time allocated for each surgeon or surgical team. The objective function is the weighted sum of the OR capacity needed and the peak bed demands. They also took into consideration downstream constraints such as hospital beds. They used probabilistic constraints in order to guarantee a probability level of the utilization of overtime in their model since they were assuming random surgery durations. They developed an approximation method based on column generation and tested with real data from the Erasmus Medical Center, in the Netherlands. In the computational study they discovered what problem parameters affect the processing time. The parameters that affect the processing time were the length of the horizon of the master plan, numbers of ORs, and number of different hospital bed types.

In [18] the authors developed a model to assign a set of elective surgeries to different ORs in each period. They assumed that the duration of the surgeries follows known distributions and also considered the random arrival of emergencies. Specifically, they defined a single random variable that represented the total OR time needed each day for emergency patients. The objective function penalized both waiting time for elective surgeries and OR overtime. They used Monte Carlo sampling techniques to approximate the stochastic component of the problem. In order to overcome the difficulty of this their problem they first reformulated the problem in a column generation scheme. Next they solved the LP relaxation of the resultant master problem, then they developed an integer feasible solution using heuristic methods (rounding and progressive reassignment) and finally they tried to improve the current solution utilizing local optimization. They provided a brief computational experiment which demonstrated the performance of the proposed algorithm considering 100 scenarios. They pointed out that more research is needed to estimate the cost of delaying elective surgeries.

In [19] surgery times are assumed deterministic but the amount of OR time needed for attending emergency cases is assumed random. The objective function is the same than [18]. The authors provided a Monte Carlo simulation approach to compute expectations of overtime and also they solve an MIP to find the master surgical plan. Since they only assume randomness in the emergency capacity needed for every period they were able to solve the approximate problem using a commercial solver. They presented computational experience for different number of scenarios from 2 to 1,000.

4.3 Operational Decisions in ORs

Once the master surgery schedule is defined, hospital management next must handle day to day operations (similar problem is found in open-block scheduling). Once the blocks of surgeries to be performed have been determined, one must next assign blocks to ORs, and schedule the surgeries withn each block. In scheduling surgeries inside of a block, one can consider both the sequence in which the surgeries are to be performed, and the scheduled starting times of each surgery. In [8] the authors model the sequencing and scheduling problem as an SIP. They measured the schedule performance using an objective function with three components: patient waiting time, OR idle time, and overtime. They tested simple rules to sequence the surgeries including: sort by variance of duration, sort by mean of duration, and sort by coefficient of variation of duration. Based on computational experience based from real data they showed that significant benefits result from sequencing and schedule the surgeries. They also found that the most robust rule was sort by variance. In [16] the authors worked on the same problem proposed in [8]. They developed an approximation algorithm based on Benders' decomposition to solve the stochastic sequencing and scheduling problem. Computational experience based on real data showed that they improved the results obtained in [8] by 15% on average. They further proved that the sample average approximation of the problem was NP-Complete.

In block scheduling situations, once each surgical block has been filled with a set of surgeries, it remains to assign the blocks to operating rooms. In [7] the authors modeled this problem as a stochastic bin packing problem. The objective function is to minimize overtime and the number of ORs opened. They proposed a set of inequalities that break the problem symmetry and also proposed different heuristics that generate good solutions in a short period of time. The computational experience shows that, on average, the simple LPT heuristic outperformed other methods proposed.

In most block scheduling problems, the surgeon is assumed to perform all his/her procedures in the same OR. In open block scheduling and other situations, the surgeon is allowed to move between ORs. When this is allowed, one must take into account that the surgeon is not required to be present for the entire procedure. It is common to break up a surgical procedure into three components: setup, cleanup, and surgery. By definition, the surgeon is only required to be present during the surgery component. If restricted to a single OR, the surgeon will be idle during the setup and cleanup phases. When the surgeon is allowed to move between ORs, there is an opportunity for increased efficiency.

In [17] the authors consider the case where one surgeon has booked two parallel ORs. Booking two ORs for a single surgeon is not uncommon when surgeries are short such as in eye surgery or orthopedics. In [17] the authors consider an objective function that included doctor waiting time, OR idle time, and overtime. The problem entails finding the sequence in which to perform the surgeries as well as the sequence in which the surgeon visits ORs (the "OR-sequence"). They modeled the problem as an SIP using sample average approximation and proposed a decomposition based solution approach based on OR-sequences. They performed a computational study to test their approach finding practical solution times. The objective function considers the tradeoff between surgeon waiting time and OR idle time. The relative cost ratio between these objective function components was estimated using a method based on creating upper and lower bounds from historical scheduling behavior. Finally, they provided a managerial insight as to when it is cost effective to assign two ORs to one surgeon.

In [2] the authors allow surgeons to move between ORs, and consider scheduling more than one surgeon in multiple ORs. They assumed that the sequence of surgeries for each surgeon is given ahead of time and then find the optimal sequence of surgeries in each OR. They used the L-shaped method in combination with a new set of cuts based on the mean value problem. They provided computational experience based on real data from the Mayo Clinic. They showed the advantage of pooling ORs as compared to assigning one OR to each surgeon. They computed the total

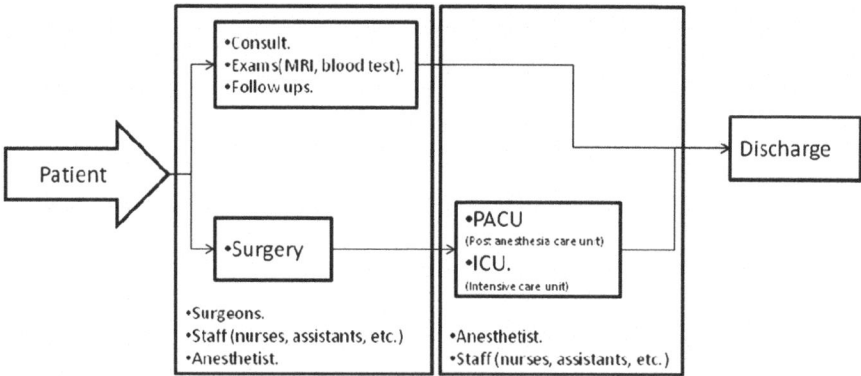

Fig. 2 Patient flow within a hospital

OR pooling cost reductions based on their estimation of cost parameters and they found a reduction of 21.78% and 58.65% on average.

4.4 Downstream Constraints

As shown in Fig. 2, once surgery is complete, most patients next go to post anesthesia care unit (PACU) to recover from anesthesia. According to [13], 15% of surgery cancelations are caused by the lack of available (PACU) beds in the hospital they studied. Therefore the construction of OR schedules that consider the downstream capacity constraints, particularly beds in the PACU and ICU, is important in hospitals where these resources are limited. Since this seems to be the case in many hospitals, this research topic is of significant interest.

In [6] the authors consider uncertainty caused by surgery duration, emergency arrivals, and turnaround times (cleanup time and setup time). They consider the problem of scheduling a set of patients over a given horizon T taking into account a finite capacity of OR time and also the number of recovery beds. They used sample average approximation to model the expected value of patient waiting time for surgery and OR overtime. They found a convenient structure in their recourse function, namely that the feasible region that defines the integer second stage region is totally uni-modular. As a result they were able to relax the integrality constraints. Unfortunately, they did not have access to historical data on ICU occupancy, so they performed a simulated experiment which showed the effect of the number of ICU beds on the solution.

Assuming that the hospital has all the resources needed in every OR to perform all surgeries is unrealistic. In fact, must facilities share special equipment among the different ORs. In [15] the authors model this situation as a resource-constrained project scheduling problem under uncertainty. They created a two-stage stochastic integer program where the random component is the time the special equipment is required within each surgery as well as the overall duration of each surgery. They used a time-index SIP formulation to model the sharing of resources. They solved the problem using Benders' decomposition and they also proposed different enhancements to speed up the processing time. They performed a computational study based on random generated instances.

4.5 Nurse Assignment

At the beginning of every shift on a hospital unit, a charge nurse must assign each nurse to a specific set of patients. This assignment determines the nurse's workload. One of the challenges of this assignment is to balance the workload since excess workload has been shown to significantly diminish the quality care of the patient. The aim of the nurse assignment problem is to balance in the workload of the nurses on the unit. The random component of this problem is the amount of time required to take care of the patient, and depends on the patient's medical condition. In [21] a two-stage stochastic integer programming, formulation is presented for the nurse assignment problem. They partitioned the shift into T periods and they penalized the amount of workload with a non-decreasing function (convex piecewise linear function in the second stage). Therefore, the marginal penalty for assigning more patient care to an overworked nurse is greater than that of a nurse with less workload. They utilized Benders' decomposition to find good solutions for a given amount of processing time (30 min). They also proposed a set of cuts to break symmetry in the master problem and also found a polynomial algorithm to solve the sub-problems. They provided computational experience based on data from the Baylor Regional Medical Center in Grapevine, Texas. They showed that their approach helped to reduce excess workload. They also mentioned that the selection of nurses in every shift has a dramatic impact. For instance, a good set of nurses might reduce the necessity of perfect information in their problem because a good nurse might spend less time with a patient and also will have less variability in patient care.

5 Open Challenges

The open questions that we have found in our study are: (1) lack of consensus about how to determine objective function costs for the multiple objective function components common in health care, (2) few if any studies modeled the dynamic nature of most healthcare operations and decisions, and (3) not many studies have developed methodology to allocate surgeries in a open block fashion.

The presence of multiple constituents in many healthcare problems (patients, physicians, hospital administration, hospital staff, etc.) commonly results in objective functions with multiple components. We have seen examples above were patient waiting time, physician idle time, staff idle time, and overtime pay are all effected by operational decisions. The relative costs of these objective function components must be estimated with reasonable accuracy before optimization techniques can be implemented. Unfortunately estimating these costs remains a very important issue in healthcare management. In [2, 8] the authors have tried to estimate these penalties through surveys or asking expert opinions. Unfortunately, we could not find a methodology that addresses this issue in detail. It would be a great benefit to have methods that help a hospital estimate these costs.

To the best of our knowledge there are no models that take into consideration the dynamics of healthcare operations. In point of fact, these operations are highly dynamic in nature. The used of multi-stage stochastic integer programming might provide a suitable approach for some of these problems. There is a great potential for methods that deal directly with both dynamic decision making and uncertainty in hospital management. However, multi-stage stochastic integer programming problems are notoriously difficult to solve. Future advances in this area would be of great interest in healthcare operations.

6 Conclusions

In this chapter we have reviewed the uses of stochastic integer programming in addressing current issues in healthcare delivery. Examples included: ambulance allocation, healthcare facility location, operating room scheduling, and nurse assignment. There remain many operational and tactical problems that seem amenable to stochastic integer programming approaches. Examples include: surgery assignment in open block scheduling, rescheduling methods to overcome disruptions to the operation room schedule, capacity expansion in health care, scheduling operations considering downstream constraints, and how to collaborate between hospitals to better serve communities.

After reviewing healthcare problems addressed using stochastic integer programming we think that this modeling technique is well suited for this application. It is able to account for uncertainty and logic decisions. Stochastic integer programming is one of the most challenging branches of optimization, and the efficiency of healthcare delivery is one of our most pressing problems. Therefore this to be a fertile area for future research.

References

1. Achieving operating room efficiency through process integration. Technical report, Health Care Financial Management Association Report (2005)
2. Batun, S., Denton, B.T., Huschka, T.R., Schaefer, A.J., The Benefit of Pooling Operating Rooms Under Uncertainty, INFORMS Journal on Computing, **23**(2), 220–237 (2011)
3. Brotcorne, L., Laporte, G., Semet, F.: Ambulance location and relocation models. Eur. J. Oper. Res. **147**(3), 451–463 (2003)

4. Caroe, C.C., Schultz, R.: Dual decomposition in stochastic integer programming. Oper. Res. Lett. **24**, 37–45 (1999)
5. Daskin, M.S., Dean, L.K.: In: Brandeau M.L., Sainfort F., Pierskalla W.P. (eds.) Location of Healthcare Facilities. Operations Research and Health Care: A Handbook of Methods and Applications, pp. 43–76. Kluwer Academic Publishers, USA (2004).
6. Daiki, M., Yuehwern, Y.: Scheduling elective surgery patients under uncertainty and downstream capacity constraints. Eur. J. Oper. Res. **206**, 642–652 (2010)
7. Denton, B.T., Miller, A., Balasubramanian, H., Huschka, T., Optimal Allocation of Surgery Blocks to Operating Rooms Under Uncertainty, Operations Research **58**(4), 802–816 (2010).
8. Denton, B.T., Viapiano, J., Vogl, A.: Optimization of surgery sequencing and scheduling decisions under uncertainty. Health Care Manage. Sci. **10**(1), 13–24 (2007)
9. Doerner, K.F., Gutjahr, W.J., Hartl, R.F., Karall, M., Reimann, M.: Heuristic solution of an extended double-coverage ambulance location problem for Austria. Central Eur. J. Oper. Res. **13**, 325–340 (2005)
10. Erdogan, S.A., Denton, B.T.: Surgery planning and scheduling: a literature review, to be submitted to Wiley Encyclopedia of Operations Research and Management Science, March (2009)
11. Erdoğan, G., Erkut, E., Ingolfsson, A., Laporte, G.: Scheduling ambulance crews for maximum coverage. J. Oper. Res. Soc. **61**, 543–550 (2010)
12. Ingolfsson, A., Budge, S., Erkut, E.: Optimal ambulance location with random delays and travel times. Health Care Manage. Sci. **11**, 262–274 (2008)
13. Jonnalagadda, R., Walrond, E.R., Hariharan, S., Walrond, M., Prasad, C.: Evaluation of the reasons for cancellation and delays of surgical procedures in a developing country. Int. J. Clin. Pract. **59**(6), 716–720 (2005)
14. Kall, P., Wallace, S.W.: Stochastic Programming. Wiley-Interscience, New York (1994)
15. Keller, B.D., Bayraksan, G.: Scheduling jobs sharing multiple resources under uncertainty: a stochastic programming approach. IIE Trans. Oper. Eng. **42**(1), 16–30 (2010)
16. Mancilla, C., Storer, R.H.: A sample average approximation approach to stochastic appointment sequencing and scheduling, IIE Transactions, **44**(8), 655–670 (2012)
17. Mancilla, C., Storer, R.H.: Stochastic sequencing of surgeries for a single surgeon operating in parallel operating rooms, technical report Lehigh University (2010)
18. Lamiri, M., Dreo, J., Xie, X.: Operating Room Planning with Random Surgery Times, Proceedings of the 3rd Annual IEEE Conference on Automation Science and Engineering, Scottsdale, AZ, USA (2007)
19. Lamiri, M., Xie, X.L., Dolgui, A., Grimaud, F.: A stochastic model for operating room planning with elective and emergency demand for surgery. Eur. J. Oper. Res. **185**, 1026–1037 (2008)
20. Özaltın, O., Johnson, M., Schaefer, A.: Senior Center Network Redesign Under Demand Uncertainty, under review
21. Punnakitikashem, P., Rosenberger, J.M., Behan, D.B.: Stochastic programming for nurse assignment. Comput. Optim. Appl. **40**(3), 321–349 (2008)
22. Shapiro, A., Dentcheva, D., Ruszczyński, A.: Lectures on Stochastic Programming: Modeling and Theory. To be published by SIAM, Philadelphia (2009)
23. Van Oostrum, J.M., Van Houdenhoven, M., Hurink, J.L., Hans, E.W., Wullink, G., Kazemier, G.: A master surgery scheduling approach for cyclic scheduling in operating room departments. OR Spectrum **30**(2), 355–374 (2008)

An Expository Discourse of E-Health

Anastasius Moumtzoglou and Anastasia Kastania

1 Introduction

Population health patterns alter the disease burden, while a higher level of education and increased availability of information raise expectations of healthcare delivery. Therefore, changes in healthcare delivery have become so widespread and numerous that the idea of e-health has become one of excitement and prediction rather than intervention [1]. Consumers, providers, organizations, and societies have seen changes in the definitions of fundamental concepts related to healthcare delivery.

On the other hand, the endorsement e-health is spreading slowly. Few companies focus on population-oriented e-health tools partly because of perceptions about the viability and capacity of the market. Moreover, developers of e-health resources are a highly diverse group with differing skills and resources while a common problem for developers is finding the balance between risk and outcome [2]. According to recent surveys, one of the most powerful restraining factors for the proliferation of e-health is the lack of security measures.

Moreover, e-health presents risks to patient health information that involve not only appropriate protocols but also laws, regulations, and appropriate safety culture. Breaches of network security and international viruses have elevated the public awareness of online information and computer security, although the over-whelming majority of security breaches do not directly involve health-related data.

Finally, as we believe in the implications of the genetic components of disease, we expect a significant increase in the genetic information of clinical

---

A. Moumtzoglou (✉)
Hellenic Society for Quality and Safety in Healthcare,
75 Evlabias Street, 13123 Ilion, Greece
e-mail: anas1@hol.gr

A. Kastania
Athens University of Economics and Business,
76 Patision Street, 10434 Athens, Greece
e-mail: ank@aueb.gr

records. Therefore, secure e-health requires not only national standardization of professional training and protocols but also interoperability of regulations and laws. As a result, professional health information organizations must take the lead in professional certification, security protocols, and applicable codes of ethics on a global basis [3].

Thus, the perspective of the chapter is to expound the idea of e-health clarifying its linkage with quality, patient safety, patient centeredness, and education.

2 Background

The term e-health has evolved through prolonged periods of time, which include the period of discovery (1989–1999), acceptance (1999–2009), and deployment (2009). Each period has unique features and applications [2]. For example, in the age of discovery, we denounced traditional approaches to understand during the era of acceptance that we need a collaboration model. Key application areas of this period include electronic medical records (including patient records, clinical management systems, digital imaging and archiving systems, e-prescribing, e-booking), telemedicine and telecare services, health information networks, decision support tools, and Internet-based technologies and services. E-health also covers virtual reality, robotics, multimedia, digital imaging, computer-assisted surgery, wearable and portable monitoring systems, and health portals. Finally, the distinctive features of the deployment period include community care, evidence-based medicine, collaborative care, and self-management.

However, the term e-health is a common neologism, which lacks precise definition. The European Commission defines e-health as "the use of modern information and communication technologies to meet needs of citizens, patients, healthcare professionals, healthcare providers, as well as policy makers." The World Health Organization offers a more precise definition by defining e-health as "the cost-effective and secure use of information and communications technologies in support of health and health-related fields, including healthcare services, health surveillance, health literature, and health education, knowledge and research." Eysenbach [4], in the most frequently cited definition, defines e-health as "an emerging field in the intersection of medical informatics, public health and business, referring to health services and information delivered or enhanced through the Internet and related technologies. In a broader sense, the term characterizes not only a technical development, but also a state-of-mind, a way of thinking, an attitude, and a commitment for networked, global thinking, to improve health care locally, regionally, and worldwide by using information and communication technology." Pagliari et al. [5] in a highly detailed analysis provide a definition, which covers individual and organizational factors. Finally, Harrison and Lee [6] provide a similar definition with a leaning towards the structural aspects of e-health applications. They refer to e-health as "an all encompassing term for the combined use of electronic information and communication technology in the health sector."

We might also describe e-health as the tools that accelerate the processing, transferring, and sharing of information between citizens, patients, and health professional [7]. These tools include health information Web sites, electronic health records, booking systems, digital image capture and sharing systems, bio-data sensors and captors, or any other of the extensive array of applications. Overall, e-health means new electronic technologies, Web-based transactions and advanced networks, and implies fundamental rethinking of healthcare processes based on electronic communication and computer-based support [8]. However, e-health does not imply a single application, as we often portray it as comprising four pillars [9]:

- Clinical applications, which include teleconsultations and clinical decision-making support software.
- E-dissemination of healthcare professional education.
- Public health information, which focuses on improving health literacy.
- Lifetime health records, which require recording and handling of information on different levels.

In conclusion, the concept of e-health is one of the recent concepts to emerge in the healthcare sector, which embodies a combination of ideas and theoretical approaches. Although the literature provides no definitional consensus on the concept, we can learn a lot about the implications of e-health through the existing literature that focuses on innovations, information technology, quality, patient safety, patient centeredness, and healthcare education. These subjects clarify the growth and impact of e-health as it relates to theoretical frameworks of systems and diffusion of innovation [1].

3 Telemedicine

Telemedicine handles both chronic and emergency medical incidences in rural and urban settings. Medical teleconsultations for telediagnosis and teletreatment allow initial diagnosis or second opinion. Nowadays, there are various asynchronous and real-time telemedicine applications including teleradiology, teledermatology, telepsychiatry, telepathology, telecytology, teleoncology, telecardiology, telesurgery, telepediatrics, home telecare, primary care, etc. Each application must meet minimal technical requirements [10] and must comply with international standards [11].

There are also various portable medical devices and smart medical sensors in the market that allow connection and transmission of biomedical signals and images through smart phones, iPad, PDAs, netbooks, and desktop personal computers using wireless or wired network infrastructures or satellite. The future vision is mobile-personalized telemedicine services [12] with emphasis to the quality and reliability of services to ensure patient safety [13] and quality of medical care at a distance [14].

Finally, there are evaluation frameworks for telemedicine systems and services [15] and guidelines [16–18] involving critical success factors towards designing and implementing national telemedicine prototypes. Security standards [11] that allow effective provision of trans-border telemedicine and telecare are also available.

4 Personalized E-Health

Mobile health applications introduce new mobile services in health care based on the technologies 2.5 (GPRS) and 3G (UMTS). This can be achieved by integrating sensors and activators to measure and transmit vital signs, with sound and video, to suppliers of health care in a Wireless Body Area Network (BAN) [19]. These sensors and the activators are improving the condition of patients while introducing new services in disease prevention and diagnostic, remote support, recording of normal conditions, and the clinical research. The Wireless Body Area Network supports the secure and reliable implementation of distant-aid services for sending reliable information from the location of accidents. Finally, the ubiquitous computing is a promising framework for creating information systems. Databases improve data management for the patients, public health, drugstores, and the workforce [20].

The field of mobile health care requires solid planning, solutions, decision-making, and knowledgeable support technology that will reduce both the number and range of faults. In a wireless world, the mobile devices use different networking installations. The existing context-aware applications are benefited from the user context (e.g., positioning information). Nevertheless, few question the network quality of service [21–23]. However, the services of electronic health share several distinctive features concerning the structure of services, the component services, and the data requirements. Therefore, the ideal is to develop electronic health services in a common platform using standard features and services.

Portable wireless applications at the point of care, based on information and connected with the clinical data, can reduce these problems. Interoperability of the existing applications of health care and/or databases is essential. The mobile technology of health care has the answer in supporting health care as well as the management of mobile patients by reducing faults and delays [24].

5 Service-Oriented Computing, Grid Computing, and Cloud Computing

Service-Oriented Computing (SOC) is the computing paradigm, which develops applications utilizing services as fundamental elements. Services are self-describing computational elements that support rapid, low-cost development of distributed applications. Services perform functions, which enable organizations to expose their core competencies over the Internet using XML-based languages and protocols. However, the basic Service-Oriented Architecture (SOA) does not address overarching concerns such as running, service orchestration, service performance management and coordination, and security.

Grid computing is a term that involves a "hot" model of distributed computing. This model embraces different architectures and forms but is restricted by specific operational and behavior rules. Its precise meaning is a difficult task because of the subject nature and the large number of its different implementations. Health Grids are computing environments of distributed resources in which diverse and scattered biomedical data are accessed, and knowledge discovery and computations are performed. The growth of the information technology allows the biomedical researchers to capitalize in ubiquitous and transparent distributed systems and in a broad assortment of tools for resource distribution (computation, storage, data, replication, security, semantic interoperability, and distribution of software as services) [25]. Electronic health records in collaboration with the Grid technologies support advanced scientific breakthrough. Finally, Medical Informatics based on a Grid can support the full spectrum of health care: screening, diagnosis, treatment planning, epidemiology, and public health. Therefore, the Grid technology supports healthcare services integration, although it is unavailable at the point of care (e.g., home care) [26].

Cloud computing encompasses different shades of meaning, which revolve around the objective of providing specific, on-demand services to multiple customers. The National Institute of Standards and Technology defines cloud computing as "a model for enabling convenient, on-demand network access to a shared pool of configurable computing resources (e.g., networks, servers, storage, applications, and services) that can be rapidly provisioned and released with minimal management effort or service provider interaction." In many firms, the cloud is the shorthand for the company's central Web servers and the services they provide. The services are categorized in the categories of Infrastructure as Service, Platform as Service, and Software as Service. Cloud computing is a natural evolution of the widespread adoption of virtualization, service-oriented architecture, autonomic, and utility computing. A service in the cloud has three distinct characteristics, which involve the sale after application, flexibility of use, and autonomous control by the supplier. Consequently, cloud computing is a technology that uses the Internet and central remote computers to protect data. Therefore, it allows efficient computing through the strengthening of storage, memory, analysis, and processes. However, cloud computing emphasizes data security because the standard encryption methods cannot be quickly adopted.

6 Quality Linkages of E-Health

Understanding the linkages between e-health and quality is a complicated task because no definitional consensus exists for either of these terms. The lack of shared definition can present challenges from both practical and academic perspectives [1].

Academically, the lack of shared definition can serve as a motivating force for a meaningful dialogue to promote knowledge. From a pragmatic point of view, different conceptualizations held by different stakeholders can also lead to serious dialogue with the intention of arriving at definitional resolution. On the other hand, the lack of consensus can cause inter- and intraorganizational dysfunctions as organizations adhere to and try to leverage their own positions for self-gain [1]. Establishing a common language may serve as a powerful platform that enables e-health to more effectively enable organizations and communities to achieve widely established goals, like quality improvement.

What is also clear about health care is that the interdependent and complex nature of healthcare delivery calls for structured ways to analyze the strategic goals, processes, technologies, outcomes, and other features of the healthcare system. Systems theory is a theoretical framework that serves as a starting point for analysis [27, 28]. This particular framework underscores the complexities of health care, but also provides a straightforward and powerful way of evaluating key relationships. As Ginter, Swayne, and Duncan show, use of systems theory allows managers and leaders to focus on the most relevant aspects of a particular problem while retaining sight of the larger context in which the issue or challenge presents itself.

In its essence, systems theory provides a straightforward way of viewing the relationship between inputs and outputs. In its most basic form, system inputs are converted to and drive outputs, which in turn provide a feedback mechanism to the size and amount of inputs needed. In an open system design, the whole system is monitored and influenced by an external environment, which may alter the nature of the system [1, 27, 28]. Austin and Boxerman's use of the cybernetic system framework is a natural extension of the systems model and provides a way in which to assess the concept of e-health as it relates to the attributes of the healthcare system, education, and quality.

Donabedian's [29] well-known structure-process-outcome (SPO) framework provides a meaningful way of viewing the concept of quality and can provide a way in which healthcare organizations can engage in critical discussions about the specific roles that e-health applications can play in improving healthcare delivery. Structural measures of quality are those that are considered "input measures of an organization's capacity to permit or promote effective work" [30], and e-health technologies and supporting equipment clearly fall under this quality domain. Because of interdependence between the quality domains under the SPO model and because of focus and importance of processes and outcomes in health care, the value of e-health applications may ultimately depend upon the ability to make significant improvements in these two dimensions.

In a systematic review of the relationship between health information technology and quality, Chaudhry et al. [31] found two main topics about this relationship. First, health information technology has been shown to increase adherence with clinical guidelines. Supporting adherence is based upon the associated decision-making processes and functions are inherently built into adherence. Second, technologies have also increased the ability of organizations to improve the quality of care by increasing monitoring through "large scale screening and aggregation of data."

On the other hand, safety culture, a macro issue, requires complete attention [2]. Safety culture is a subset of the organizational culture [32]. Although an overriding concern in recent years, we need further work to identify and soundly determine the key dimensions of patient safety culture [33, 34] and understand its relationship with the leadership. A prerequisite for the realization of this perspective is the collection, analysis, and dissemination of information deriving from incidents and near misses as well as the adoption of the reporting, just, flexible, and learning cultures [35]. As patients become increasingly concerned about safety, clinical medicine focuses more on quality [36] than on bedside teaching and education.

To substantiate the argument that e-health is the future realm of healthcare quality [2], we identify emerging scientific fields, and evaluate their impact on quality and patient safety:

- Biobanking is an emerging discipline [37], which follows the continuing development of new techniques and goals. Overall, it constitutes a mechanism, which facilitates the understanding of the genetic basis of disease [38] and holds taxonomic strains [39]. However, its operation and maintenance is a skill-rich activity, which requires careful attention to the implementation of preservation technologies. Moreover, biobanking represents a challenge to informed consent [38] while only appropriate quality assurance ensures that recovered cultures and other biological materials behave in the same way as the originally isolated culture.
- A biochip is a set of microarrays arranged on a solid substrate which permits to perform various tests at the same time [40, 41]. It replaces the standard reaction platform and produces a patient profile, which we use in disease screening, diagnosis, and monitoring disease progression. The development of biochips is a substantial thrust of the rapidly growing biotechnology industry, which encompasses a diverse range of research efforts.
- Data mining is the process of extracting patterns from data. It supports workflow analysis [42] and saves time and energy leading to less hassle for clinicians. While data mining can be used to identify patterns in data samples, it is essential to understand that non-representative samples may cause non-indicative results. Data mining is becoming an increasingly indispensable tool in mining a wide range of health records [43]. Moreover, we might accompany it with semantics-based reasoning in the management of medicines. Finally, data mining can maintain quality assurance [44], simplify the automation of data retrieval, facilitate physician quality improvement [45], and accurately represent patient outcomes if combined with simulation [46]. Recently, there is interest in switching to algorithms and database development for microarray data mining [47].
- Disease modeling, the mathematical representation of a clinical condition, summarizes the understanding of disease epidemiology, and requires computational modeling, which follows two different approaches.
- Genomics, the study of the genomes of living entities, encompasses considerable efforts to understand the complete DNA sequence of organisms through fine-scale genetic mapping efforts. It also includes studies of intragenomic phenomena

and interactions between loci and alleles within the genome. Overall, many disciplines turn on the issue of automating the different stages in post-genomic research with a view to developing high-dimensional data of high quality [48].
- Molecular Imaging unites molecular biology and in vivo imaging while enabling the visualization of the cellular function and the follow-up of the molecular process [49]. Molecular imaging imparts a greater degree of fairness to quantitative tests, and numerous potentialities to the diagnosis of cancer, neurological and cardiovascular diseases. Therefore, it has a significant economic impact due to earlier and more accurate diagnosis.
- Nanotechnology encompasses the extension of traditional device physics, different approaches based upon molecular self-assembly, and new materials with nanoscale dimensions. Nanomedicine, the medical use of nanotechnology, encompasses the use of nanomaterials and nanoelectronic biosensors and seeks to provide a valuable body of research and tools [50, 51]. New applications in the pharmaceutical industry include advanced drug delivery systems, alternative therapies, and in vivo imaging. Moreover, molecular nanotechnology, a preliminary subfield of nanotechnology, deals with the engineering of molecular assemblers. Finally, neuro-electronic interfacing, an innovative project dealing with the creation of nanodevices, will connect computers to the nervous system while nanonephrology will play a role in the management of patients with kidney disease.
- Ontologies have become a mainstream issue in biomedical research [52] since we can explain biological entities by using annotations. This type of comparability, which we call semantic similarity, assesses the extent of connectedness between two entities using annotations similarity.
- Proteomics, a term coined in 1997 to make an analogy with genomics, is the detailed study of proteins, their structure and function [53, 54]. One of the most promising developments from the study of human genes and proteins is the discovery of new drugs. This relies on the identification of proteins associated with a disease, and involves computer software which uses proteins as targets for new drugs.
- Medical simulation bridges the knowledge gap by representing certain key characteristics of a physical system. Quality improvement, patient safety, and the evaluation of clinical skills have impelled medical simulation into the clinical arena [55]. Still, there is conclusive evidence that simulation training improves provider self-efficacy and effectiveness [56] and increases patient safety. Finally, the process of iterative learning creates a much stronger learning environment and computer simulators are an ideal tool for evaluation of students' clinical skills [57].

7 Patient Centered Care Linkages of E-Health

Kilbridge [58] suggested that 12 information technology applications, which involve the management of health care information, empower patients: (1) technologies which provide access to general and specific health care information and (2) technologies capable to handle data entry and tracking of individual self-management data. The applications included:

Personal Health Records: The Personal Health Record contributes to patient empowerment by granting patients access to their medical record and permitting them to explore and, modify that information [59–66].

Patient Access to Hospital Information Systems: A subcategory of the personal health record is a technology that permits outpatients to gain access to their laboratory test results, imaging research reports, etc. through online access to hospital information systems.

Patient Access to General Health Information: Patient Access to General Health Information can empower patients in discussing treatment with providers and, in particular, alter the balance of power in the provider–patient relationship.

Electronic Medical Records (EMRs): The sophisticated EMRs preserve reliability because they provide clinical decision support functions, especially the capacity to influence adherence to guidelines in diagnosis, treatment, and prescribing.

Pre-Visit Intake: These applications allow patients to do a self-assessment and create a health profile for presentation to their physician.

Inter-Hospital Data Sharing: Central to these schemes is the capacity for remote access to hospital clinical information, from departmental systems (laboratory, radiology, etc.) or from a central clinical data repository. The technologies are either Internet-based or rely upon dependable dial-up connections and increase efficiency in various ways.

Information for Physicians to Manage Patient Populations: These technologies assist providers in tracking and managing populations of patients, according to clinical practice guidelines.

Patient–Physician Electronic Messaging: Electronic messaging is a self-documenting improvement over the phone contact [67–71].

Patient Access to Tailored Medical Information, Online Data Entry, and Tracking: These applications address patients' medical condition in the context of their personal clinical data. They include online "disease management" applications, apply online data-entry and the results can be made available to physicians to assist them in case monitoring. These technologies support patient self-care as well as provider monitoring and support of care between office visits.

Online Scheduling: Online Scheduling provides patients the way to connect with a practice over the Internet and schedule appointments. The technology improves the physician–patient relationship by facilitating access to care, reducing the duration and inconvenience of the standard appointment scheduling practice.

Computer-Assisted Telephone Triage and Assistance: Various models of call center technology improve communications between patients and their caregivers. They represent a more convenient form of communication with the provider.

Online Access to Provider Performance Data: Governments, regulatory bodies, and private companies offer free online access to physician and hospital performance or quality data.

8 Educational Linkages of E-Health

A key way to determine e-health and educational linkages relates to the ways in which e-health applications and technologies can be introduced into educational settings. On the one hand, the technologies, applications, concepts, and ideas can be represented in programs of instruction and curricula for prospective and current providers and technicians. On the other hand, these topics could be introduced in such a way as to educate students on the mechanics of the applications—that is, the focus would shift from pure pedagogy to modeling and application. The focus here would entail hands-on learning. Visualizing the classroom and technologically advanced forms of classroom delivery as two broad categories leads to the development of a simple model to study relationships between e-health and education [1].

9 Issues, Controversies, and Problems

Financial and other barriers that result in comparatively low penetration rates of e-health remain [2]. These barriers include acquisition and implementation costs, the lack of interoperability standards, skepticism about the business case for investment, uncertainty about system longevity, and psychological barriers related to uncertainty and change. E-health systems are expensive and providers face problems making the investment. Even if they can handle the initial investment and implementation costs, they must remain confident that an e-health system will improve efficiency or cover its upfront and ongoing costs to make the investment. In addition, some providers' reluctance to embrace e-health may be a rational response to skewed financial incentives since the health care system provides little financial incentive for quality improvement [8]. Perhaps the most controversial barrier to adoption of quality-related information technology is the lack of incentives for change. Therefore, nongovernmental groups might play a highly pivotal role in developing incentives, and regulatory agencies can apply for reasonable, standardized, and more easily gathered quality measurement data. Additional barriers to the spread of e-health include legal and regulatory concerns and technological issues [72]. Legal challenges include privacy of identifiable health information, reliability and quality of health data, and tort-based liability while recommendations for legal reform include [73]:

- Sensitivity of health information
- Privacy safeguards
- Patient empowerment
- Data and security protection

Technological barriers to e-health stem from the evolutionary nature of these systems but also obsolescence, and the lack of standards or criteria for interoperability. Without the technical specifications that enable interoperability, data exchange between providers who use different e-health systems is severely limited. Moreover, security and confidentiality in information technology represent a serious matter [74]. Finally, implementation and dissemination issues are decipherable. The effects of e-health tools on patient behavior, the patient–clinician relationship, the legal and ethical implications of using health information technologies, and clinical decision support systems are unclear. Furthermore, potential health inequalities resulting from the digital divide have to keep within bounds. Overall, key questions include clinical decision support, revision of guidelines for local implementation, implementation and dissemination of clinical information systems, patient involvement, and the role of the Internet.

10 Solutions and Recommendations

Leadership should work in conjunction with the staff in order to mitigate any apprehension relating to e-health [75, 76]. With this internal support and commitment, healthcare professionals will become involved and integrate e-health into their daily practice. Moreover, there is a fundamental relationship between organizational and technological change [77], which providers should understand prior to implementation because they constitute the driving force behind changes within the clinical setting. Conclusively, the successful implementation of e-health requires thoughtful integration of standard procedures and information technology elements.

11 Conclusion

E-health, which already supports significant advances in healthcare quality, patient safety, and patient centeredness, has the potential to become their realm [2], if we develop an outline of the cognitive and social factors related to their design and use and integrate them with the clinical practice. E-health has an enormous potential to empower citizens, patients, and healthcare professionals. It can also provide governments a way to cope with increasing demand for healthcare services and reshape the expectations of healthcare delivery.

References

1. Pate, C., Turner-Ferrier, J.: Expoloring linkages between Quality, E-health and Healthcare Education. In: Kastania, A., Moumtzoglou, A. (eds.) E-Health Systems Quality and Reliability: Models and Standards. IGI Global, Hershey, PA (2010)
2. Moumtzoglou, A.: E-health as the realm of healthcare quality: a mental image of the future. In: Kastania, A., Moumtzoglou, A. (eds.) E-Health Systems Quality and Reliability: Models and Standards. IGI Global, Hershey, PA (2010)

3. Kluge, E.H.: Secure e-health: managing risks to patient health data. Int. J. Med. Informat. **76**(5–6), 402–406 (2007)
4. Eysenbach, G.: What is e-health? J. Med. Internet Res. **3**(2), e20 (2001)
5. Pagliari, C., Sloan, D., Gregor, P., et al. What is eHealth?: a scoping exercise to map the field. J. Med. Internet Res. 31:7(1):e9 (2005)
6. Harrison, J., Lee, A.: The role of e-health in the changing health care environment. Nurs. Econ. **24**(6), 283–289 (2006)
7. Wilson, P.: E-health—building on strength to provide better healthcare anytime anywhere. Paper presented at the eHealth 2005 Conference, Tromsø, Norway (2005)
8. Blumenthal, D., Glaser, J.: Information technology comes to medicine. New Engl. J. Med. **356**(24), 2527–34 (2007)
9. Richardson, R., Schug, S., Bywater, M., Lloyd-Williams, D.: Development of eHealth in Europe: Position Paper. European Health Telematics Association, Brussels (2004)
10. Ferrer-Roca, O., Iuindicissa, M.S.: Handbook of Telemedicine. Studies in Health Technology and Informatics, vol. 54, IOS Press/Ohmhsa, Amsterdam (1998)
11. Ferrer-Roca, O.: Standards in telemedicine. In: Kastania, A., Moumtzoglou, A. (eds.) E-Health Systems Quality and Reliability: Models and Standards. IGI Global, Hershey, PA (2010)
12. Kastania, A., Kossida, S.: Quality issues in personalized e-health, mobile health and e-health Grids. In: Kastania, A., Moumtzoglou, A. (eds.) E-Health Systems Quality and Reliability: Models and Standards. IGI Global, Hershey, PA (2010)
13. Zimeras, S., Kastania, A.: Statistical models for EHR security in Web healthcare information systems. In: Varlamis, I., Chryssanthou, A., Apostolakis, I. (eds.) Certification and Security in Health-Related Web Applications: Concepts and Solutions. IGI Global, Hershey, PA (2010)
14. Kastania, A., Zimeras, S.: Quality and reliability aspects in telehealth systems. In: Siassiakos, K., Lazakidou, A. (eds.) Handbook of Research on Distributed Medical Informatics and E-Health. IGI Global, Hershey, PA (2008)
15. Field, M.: Telemedicine—A Guide to Assessing Telecommunications in Health Care. Institute of Medicine, National Academy Press, Atlanta (1996)
16. World Health Organization. eHealth tools and services—needs of the member states, Report of the WHO Global Observatory for eHealth, WHO Press, Geneva (2006a)
17. World Health Organization. Building Foundations for e-health: Progress of Member States, Report of the WHO Global Observatory for eHealth, WHO Press, Geneva (2006b)
18. World Health Organization. Telemedicine—Opportunities and developments in Member States, Global Observatory for eHealth series—Volume 2. http://www.who.int/goe/publications/ehealth_series_vol2/en/index.html (2011). Accessed 16 Mar 2011
19. Konstantas, D., Jones, V., Herzog, R.: MobiHealth-innovative 2.5/3G mobile services and applications for healthcare. Paper presented at the Eleventh Information Society Technologies (IST) Mobile and Wireless Telecommunications (2002)
20. Milenkovi, A., Otto, C., Jovanov, E.: Wireless sensor networks for personal health monitoring: issues and an implementation. Comp. Commun. (Special issue: Wireless Sensor Networks: Performance, Reliability, Security, and Beyond) 29(13–14):2521–2533 (2006)
21. Wac, K.: Towards QoS-awareness of context-aware mobile applications and services. Paper presented at the On the Move to Meaningful Internet Systems 2005: OTM Workshops, Ph.D. Student Symposium (2005)
22. Broens, T., van Halteren, A., van Sinderen, M., Wac, K.: Towards an application framework for context-aware m-health applications. Int. J. Internet Protocol. Technol. **2**(2), 109–116 (2007)
23. Hanak, D., Szijarto, G., Takacs, B.: A mobile approach to ambient assisted living. Paper presented at the IADIS Wireless Applications and Computing (2007)
24. Archer, N.: Mobile eHealth: Making the Case Euromgov2005: Mobile Government Consortium International. (2005)
25. Kratz, M., Silverstein, J., Dev, P. HealthGrid: Grid Technologies for Biomedicine. Telemedicine & Advanced Technology Research Center, U.S. Army Medical Research and Materiel Command, Fort Detrick, Maryland (2007)

26. Koufi, V., Malamateniou, F., Vassilacopoulos, G.: A medical diagnostic and treatment advice system for the provision of home care. Paper presented at the 1st international conference on PErvasive Technologies Related to Assistive Environments (2008)
27. Austin, C.J., Boxerman, S.B.: Information systems for health services administration, 5th edn. Health Administration Press, Chicago, IL (1998)
28. Ginter, P.M., Swayne, L.E., Duncan, W.J.: Strategic management of health care organizations, 3rd edn. Blackwell Publishers Inc., MA (1998)
29. Donabedian, A.: Evaluating the quality of medical care. Milbank Q. **83**(4), 691–729 (2005)
30. Flood, A.B., Zinn, J.S., Shortell, S.M., Scott, W.R.: Organizational performance: managing for success. In: Shortell S.M., Kaluzny A.D. (eds.) Health Care Management: Organization Design and Behavior. Delmar, NY (2000)
31. Chaudhry, B., Jerome, W., Shinyi, W., Maglione, M., Mojica, W., Roth, E., et al.: Systematic review: impact of health information technology on quality, efficiency, and costs of medical care. Ann. Intern. Med. **144**(10), E12–W18 (2006)
32. Olive, C., O'Connor, T.M., Mannan, M.S.: Relationship of safety culture and process safety. J. Hazard. Mater. **130**(1–2), 133–140 (2006)
33. Ginsburg, L., Gilin, D., Tregunno, D., Norton, P.G., Flemons, W., Fleming, M.: Advancing measurement of patient safety culture. Health Serv. Res. **44**(1), 205–224 (2009)
34. Singer, S.J., Falwell, A., Gaba, D.M., Baker, L.C.: Patient safety climate in US hospitals: variation by management level. Med. Care **46**(11), 1149–1156 (2008)
35. Ruchlin, H.S., Dubbs, N.L., Callahan, M.A.: The role of leadership in instilling a culture of safety: lessons from the literature. J. Healthc. Manag. **49**(1), 47–58 (2004)
36. Simmons, B., Wagner, S.: Assessment of continuing interprofessional education: lessons learned. J. Contin. Educ. Health Prof. **29**(3), 168–171 (2009)
37. Riegman, P.H., Morente, M.M., Betsou, F., de Blasio, P., Geary, P.: Biobanking for better healthcare. Mol. Oncol. **2**(3), 213–222 (2008)
38. Ormond, K.E., Cirino, A.L., Helenowski, I.B., Chisholm, R.L., Wolf, W.A.: Assessing the understanding of biobank participants. Am. J. Med. Genet. **149A**(2), 188–198 (2009)
39. Day, J.G., Stacey, G.N.: Biobanking. Mol. Biotechnol. **40**(2), 202–213 (2008)
40. Fan, Z.H., Das, C., Chen, H.: Two-dimensional electrophoresis in a chip. In: Keith E., Herold, K., Rasooly, A. (eds.) Lab-on-a-Chip Technology: Biomolecular Separation and Analysis. Caister Academic Press, Norfolk (2009)
41. Cady, N.C.: Microchip-based PCR Amplification Systems. Lab-on-a-Chip Technology: Biomolecular Separation and Analysis. Caister Academic Press, Portland (2009)
42. Lang, M., Kirpekar, N., Burkle, T., Laumann, S., Prokosch, H.U.: Results from data mining in a radiology department: the relevance of data quality. Stud. Health Tech. Informat. **129**(Pt 1), 576–580 (2007)
43. Norén, G.N., Bate, A., Hopstadius, J., Star, K., Edwards, I.R.: Temporal pattern discovery for trends and transient effects: its application to patient records. Proceedings of the Fourteenth International Conference on Knowledge Discovery and Data Mining SIGKDD 2008, pp 963–971. Las Vegas NV (2008)
44. Jones, A.R.: Data mining can support quality assurance. J. Roy. Soc. Med. **102**(9), 358–359 (2009)
45. Johnstone, P.A., Crenshaw, T., Cassels, D.G., Fox, T.H.: Automated data mining of a proprietary database system for physician quality improvement. Int. J. Radiat. Oncol. Biol. Phys. **70**(5), 1537–1541 (2008)
46. Harper, P.: Combining data mining tools with health care models for improved understanding of health processes and resource utilisation. Clin. Invest. Med. **28**(6), 338–341 (2005)
47. Cordero, F., Botta, M., Calogero, R.A.: Microarray data analysis and mining approaches. Brief. Funct. Genomic Proteomic. **6**(4), 265–281 (2007)
48. Laghaee, A., Malcolm, C., Hallam, J., Ghazal, P.: Artificial intelligence and robotics in high throughput post-genomics. Drug Discov. Today **10**(18), 1253–1259 (2005)
49. Weissleder, R., Mahmood, U.: Molecular imaging. Radiology **219**, 316–333 (2001)

50. Wagner, V., Dullaart, A., Bock, A.K., Zweck, A.: The emerging nanomedicine landscape. Nat. Biotechnol. **24**(10), 1211–1217 (2006)
51. Freitas Jr., R.A.: What is Nanomedicine? Nanomedicine: Nanotech. Biol. Med. **1**(1), 2–9 (2005)
52. Pesquita, C., Faria, D., Falcao, A.O., Lord, P., Couto, F.M.: Semantic similarity in biomedical ontologies. PLoS Comput. Biol. **5**(7), e1000443 (2009)
53. Anderson, N., Anderson, N.G.: Proteome and proteomics: new technologies, new concepts, and new words. Electrophoresis **19**(11), 1853–61 (1998)
54. Blackstock, W.P., Weir, M.P.: Proteomics: quantitative and physical mapping of cellular proteins. Trends. Biotechnol. **17**(3), 121–7 (1999)
55. Carroll, J.D., Messenger, J.C.: Medical simulation: the new tool for training and skill assessment. Perspect Biol. Med. **51**(1), 47–60 (2008)
56. Nishisaki, A., Keren, R., Nadkarni, V.: Does simulation improve patient safety?: self-efficacy, competence, operational performance, and patient safety. Anesthesiol. Clin. **25**(2), 225–36 (2007)
57. Murphy, D., Challacombe, B., Nedas, T., Elhage, O., Althoefer, K., Seneviratne, L., Dasgupta, P.: Equipment and technology in robotics (in Spanish; Castilian). Archivos espanoles de urologia **60**(4), 349–55 (2007)
58. Kilbridge, P.: Crossing the Chasm with Information Technology: Bridging the Quality Gap in Health Care. California HealthCare Foundation, CA (2002)
59. Halamka, J.: Making smart investments in health information technology: core principles. Health Aff. (Millwood) **28**, 385–389 (2009)
60. Hess, R., Bryce, C., Paone, S., Fischer, G., McTigue, K., Olshansky, E., Zickmund, S., Fitzgerald, K., Siminerio, L.: Exploring challenges and potentials of personal health records in diabetes self-management: implementation and initial assessment. Telemed. E Health **13**(5), 509–18 (2007)
61. Kupchunas, W.R.: Personal health record: new opportunity for patient education. Orthop. Nurs. **26**(3), 185–91 (2007)
62. Lee, M., Delaney, C., Moorhead, S.: Building a personal health record from a nursing perspective. Int. J. Med. Informat. **76**(Suppl 2), S308–S316 (2007)
63. Morales Rodriguez, M., Casper, G., Brennan, P.F.: Patient-centered design. The potential of user-centered design in personal health records. J. Am. Health Inform. Manage. Assoc. **78**(4), 44–6 (2007)
64. Nelson, R.: The personal health record. Am. J. Nurs. **107**(9), 27–8 (2007)
65. Pagliari, C., Detmer, D., Singleton, P.: Potential of electronic personal health records. Br. Med. J. **335**(7615), 330–3 (2007)
66. Smith, S.P., Barefield, A.C.: Patients meet technology: the newest in patient-centered care initiatives. Health Care Manag. (Frederick) **26**(4), 354–62 (2007)
67. Anand, S., Feldman, M., Geller, D., Bisbee, A., Bauchner, H.: A content analysis of e-mail communication between primary care providers and parents. Pediatrics **115**, 1283–1288 (2005)
68. Car, J., Sheikh, A.: Email consultations in health care: 2—acceptability and safe application. Br. Med. J. **329**, 439–442 (2004)
69. Kleiner, K., Akers, R., Burke, B., Werner, E.: Parent and physician attitudes regarding electronic communication in pediatric practices. Pediatrics **109**, 740–744 (2002)
70. Leong, S., Gingrich, D., Lewis, P., Mauger, D., George, J.: Enhancing doctor-patient communication using email: a pilot study. J. Am. Board Fam. Pract. **18**(3), 180–8 (2005)
71. Stalberg, P., Yeh, M., Ketteridge, G., Delbridge, H., Delbridge, L.: E-mail access and improved communication between patient and surgeon. Arch. Surg. **143**(2), 64–169 (2008)
72. Davenport, K.: Navigating American Health Care: How Information Technology Can Foster Health Care Improvement. Center for American Progress, http://www.americanprogress.com (2007). Accessed 24 Apr 2009
73. Hodge Jr., J.G., Gostin, L.O., Jacobson, P.D.: Legal issues concerning electronic health information: privacy, quality, and liability. J. Am. Med. Assoc. **282**(15), 1466–71 (1999)
74. Anderson, J.G.: Social, ethical and legal barriers to e-health. Int. J. Med. Informat. **76**(5–6), 480–3 (2007)

75. Iakovidis, I.: Towards personal health record: current situation, obstacles, and trends in implementation of electronic healthcare record in Europe. Int. J. Med. Informat. **52**, 105–115 (1998)
76. Burton, L.C., Anderson, G.F., Kues, I.W.: Using health records to help coordinate care. Milbank Q. **82**(3), 457–481 (2004)
77. Berg, M.: Implementing information systems in health care organizations: myths and challenges. Int. J. Med. Informat. **64**, 143–156 (2001)

Nurse Scheduling Problem: An Integer Programming Model with a Practical Application

Neng Fan, Syed Mujahid, Jicong Zhang, Pando Georgiev, Petraq Papajorgji, Ingrida Steponavice, Britta Neugaard, and Panos M. Pardalos

1 Introduction

The Nurse scheduling problem (NSP) is the problem of determining a work schedule for nurses that satisfies a set of requirements and preferences of nurses. Despite seeming trivial, this is a complex problem due to its many constraints and many possible combinations. NSP is all about assignment of shifts and leave time for nurses. The nurse manager creates a schedule considering personal preferences and/or scheduling requirements. The problem is described as finding a schedule that supports the preferences of nurses and fulfills the requirement of hospital management.

Conventionally the workday is divided into shifts that could be day, evening or night shifts. In the case considered in this study, there are seven shifts which will be classified into day/evening/night shifts, and the time lengths of these shifts are either 8.5 or 12.5 h. Each shift has its only minimum and maximum required number of

---

N. Fan
Department of Systems and Industrial Engineering, University of Arizona, AZ, USA

S. Mujahid • P. Georgiev (✉) • P. Papajorgji • P.M. Pardalos
Department of Industrial and Systems Engineering,
Center for Applied Optimization, University of Florida,
Gainesville, FL, USA
e-mail: pandogeorgiev@gmail.com

J. Zhang
Functional Neuroscience and Neurosurgery Laboratory, Department of Neurology/Neurosurgery, School of Medicine, The Johns Hopkins University, Baltimore, MD, USA

I. Steponavice
Department of Mathematical Information Technology, University of Jyvaskyla, Finland

B. Neugaard
James A. Haley VAMC, 13000 Bruce B. Downs Blvd, Tampa, FL 33612, USA

College of Public Health, University of South Florida,
Tampa, FL, USA

nurses. Besides considering the scheduling of two types of nurses [registered nurses (RN) and licensed practical nurses (LPN)], two other types of employees [nursing assistants (NA) and health care techs (HCT)] are also included in our model. Each has its own available number of employees. In addition, two employees, the Nurse Manager and the Clinical Nurse Leader, who mostly play supervising roles, are also included. Most of the employees are full-time staff. In our model, a full-time employee needs to work 80 h per 2 weeks. Some of them are working part time and a part-time staff works less than 80 h per week.

The NSP in our practice has two stages: the first one is to collect the preference of each employee (including both nurses and other staffs); the second one is to present a schedule for the all employees in a given period of time. In our case, the scheduling of employees is created for a period of 4 weeks. Thus employees are asked to make their choices of working shifts for a period of 4 weeks in advance. We mentioned above that there are four types of employees, and we consider them as four groups; each group has a ranking number that allows members of different groups to have a rotating priority while presenting their choices. Therefore, each group will have, at some point, the highest priority to select working shifts.

Within the same group, members will have a ranking system as well. The ranking within the group can be decided by the time employees enter their preferences into the system, or by an administrator who may take into account other considerations, such as personal problems, or other administrative reasons. In the event that two employees plan to work in the same shift on the same day, their rank will decide who can choose this specific shift first. The creator of the schedule will collect all of the employees' preferences before preparing the schedule. In our model we define a period to be a month (4 weeks, 28 days). In addition, an employee may work more than 6 days per week if he/she prefers to work 7 days a week. These kinds of preferences of each nurse should also be considered by the administrative staff creating the schedule. For obvious reasons, the employees may have other requirement such as: a nurse does not work the day shift, evening shift, and night shift on the same day; a nurse may go on vacation and will not work shifts during this time; a nurse does not work a late night shift followed by a day shift the next day; a nurse needs enough time off between two consecutive shifts.

The NSP is to find a schedule that both respects the preferences of the employees and fulfills the objectives of the hospital. The need of the hospital includes all minimum requirements for each shift should be satisfied. Besides the constraints of the requirements of employees and hospital, there are also many other constraints for these employees such as working hours per each week. Due to high number of constraints and the many possible solutions of NSP, this problem, like many scheduling problems, appears to be NP-hard.

The NSP been studied in many approaches, and two recent surveys discussed them [4, 5]. Mathematically, heuristic and optimization methods are most widely used. The heuristic methods include genetic or local search algorithms [3, 6, 8, 14, 16]. The commonly used optimization methods are linear programming [12], dynamic programming [11], stochastic programming [1], nonlinear programming [18], integer programming [2], and goal programming [9]. In addition, some fuzzy models can be found [15].

Differently from the previous research [4, 5, 10, 17, 19], in our model, we distinguish the schedule of weekdays and weekends with different requirements and also consider specific requirements of specified employees in some shifts in real practical applications. In addition, the seven shifts on each day do not necessarily have the same length and this is different from that in the paper [10], which considers the shifts having a length of 4 h. We also consider some part-time nurses in our problem, which is previously considered in the paper [13]. In this paper, we use binary integer linear programming to formulate the NSP with the specific requirement of a VA hospital. The objective is to maximize the satisfactions of all employees and maintain the hospital's requirements. Our algorithm is implemented in CPLEX [7], and it appears to be quite efficient for obtaining the schedule in our model.

The rest of this paper is organized as follows: Sect. 2 is the detailed description of our specific VA hospital scheduling problem; Sect. 3 is the binary integer formulation of this problem; Sect. 4 includes the implementations of the model and the results of a practical example; Sect. 5 concludes the paper.

2 The Problem Description: Goals and Requirements

2.1 Two Stages of Nurse Scheduling Problem

As described in Sect. 1, the NSP in our practice has two stages: collecting each employee's preference and presenting a schedule for all employees in a given period. The preference of an employee is defined to be her/his choice to work in which shift during the week. The rotating priority rank of each employee is chosen before scheduling and it is decided by the administrator for a rotating purpose for all employees in advance to ensure a fair assignment to all of them. In addition, each employee has to declare whether he/she prefers to work 7 days in a given week.

The second stage of presenting the schedule is defined to be decision variables for each employee corresponding to some shift of a day in a week. The schedule will be decided by solving an optimization model.

2.2 Data of Nurse Scheduling Problem

The data for scheduling listed in the following is classified into two types. The datasets in (a), (b), (c) should have no changes for a long time, for example, a year. The datasets in (c), (d), (e) should be collected at the beginning of each scheduling period, for example, a month (in the following, defined as 28 days, 4 weeks), and they change over each period (a month). The dataset (g) is chosen to satisfy the holiday requirements of nurses.

Table 1 Seven shifts

Shift number	Time period	Classify
S1	07:30–16:00	Day shift
S2	07:30–20:00	
S3	15:30–24:00	Evening shift
S4	15:30–04:00	(Over)night shift
S5	19:30–08:00	
S6	23:30–08:00	
S7	03:30–16:00	

Table 2 Four types of employees and one manager, one clinical nurse leader

Types of employees		Total nurses	Full time/part time			
			100%	90 %	60 %	30 %
t1	RN	29	26	1	1	1
t2	LPN	11	10		1	
t3/t4	NA/HCT	7	7			
t5	Nurse manager (NM)	1	1			
t6	Clinical nurse leader (NL)	1	1			
Total		47+2	43+2	1×0.9	2×0.6	1×0.3

Table 3 Requirements with respect to types and shifts for both weekdays and weekends

		Days (0730–1,600, shift 1)	Evenings (1,530–2,400, shift 3)	Nights (1,200–0800, shift 6)
RN	Max	8	8	8
	Min	3/2	3	3
LPN	Max	5	5	5
	Min	0	0	0
HCT/NA	Max	n/a	n/a	n/a
	Min	0	0	0

(a) The shifts with begin/end time are decided at the very beginning (Table 1).
(b) The number of employees in each type is given at the very beginning (Table 2).
(c) The requirement for each shift is given at the very beginning (Tables 3 and 4).
(d) The preferred scheduling (preference) of each employee is collected before each scheduling (Table 5).
(e) The preference of whether to work 7 days a week or not is collected before each scheduling (Table 5).
(f) The priority rank of each group is picked up before each scheduling (Table 5).
(g) Holiday/vacation leaves requirement.

These data should be given before presenting the schedule for the whole employees. Other data may include the emergent leaves, or specific requirements such as the casual needs of nurses in some shifts (b), (c).

Table 4 Numbers of required employees for each shift

Shift number	Weekdays		Weekends	
	Min	Max	Min	Max
S1	7	8	7	8
S2	3	4	3	4
S3	7	8	6	7
S4	1	1	1	1
S5	3	4	3	4
S6	6	7	5	6
S7	1	1	1	1

Table 5 Data with changes over time

Data	Description	
Preference	Binary	Each preference is defined to be either 1 (preferred) or 0 of the employee to work or not to work on the specific shift of some week. This can be represented in a binary tensor with four dimensions
Seven Days	Binary	The employee can choose to work 7 days a week if he/she prefers. This can be stored in a binary tensor of three dimensions
Priority Rank	Integer	The priority rank. The rotating priority rank of each employee is chosen before scheduling and it is decided by the administrator for a rotating purpose for all employees in a long time to ensure a fair assignment for all of them

In the following, the datasets mentioned above are presented in tables.

In Table 1, the length of shift 1, shift 3, and shift 6 is 8.5 h, while the length of shift 2, shift 4, shift 5, and shift 7 is 12.5 h. In addition, shift 4, shift 5, and shift 6 are overnight shifts. For each day, there are seven shifts within it.

In Table 1, we can find all five overlaps among different shifts.

0730-0800: S1, S2 begin and S5, S6 end; 1530-1600: S3, S4 begin and S1, S7 end;1930-2000: S5 begins and S2 ends; 2330-2400: S6 begins and S3 ends, and0330-0400: S7 begins and S4 ends.

In Table 2, we consider the nurse manager and the clinical nurse leader as two types of employees (t5, t6). The numbers under the cell of full time/part time denote the percentages of hours this employee should work per 2 weeks: 100 % means 80 h per 2 week, 90 % means 72 h per 2 weeks, 60 % means 48 h per 2 weeks, and 30 % means 24 h per 2 weeks.

In Table 3, all numbers are the requirement numbers of employees with respect to that shift of specific type for all weekdays and weekends except that the one 3/2 in Table 3 is considered as weekday/weekend's requirement.

The notations for these three kinds of data are in Sect. 3.1.

2.3 The Requirements of NSP: Hard and Soft Constraints

We have two types of constraints: hard constraints are defined as if this constraint fails then the entire schedule is invalid; soft constraints are desirable that these constraints are met but not meeting them does not make the schedule invalid. The following constraints are considered as the hard constraints since they are all required by the hospital. The satisfactions of pre-chosen preferences for all employees are considered as the soft constraints. For our application in a VA hospital, we have the hard constraints as follows

1. The number of employees for each shift should be satisfied;
2. RN may not work more than 12 h per shift, and LPN, NA, HCT may not work greater than 16 h per week;
3. RN may not work more than 60 h per week;
4. RN, LPN, NA, HCT may not work more than 6 days a week if he/she does not prefer to work 7 days in that week;
5. RN may not work more than 7 days consecutively;
6. All employees should work 4–6 weekend days per month;
7. The schedule of an employee should avoid shift overlap contradiction and also guarantee enough rest time between two consecutive shifts of that employee (Table 6);
8. The different time requirement for full-time and part-time employees should be satisfied;
9. The manager and the clinical nurse leader present day shift only;
10. Each employee works at most one shift per day;
11. Each employee in 1 week works only on one shift;
12. Some employees may need vacation leaves.

Table 6 Rest time between two consecutive shifts

Shifts		S1	S2	S3	S4	S5	S6	S7
	S1	1→1	1→2	1→3	1→4	1→5	1→6	1→7
Rest hrs		15.5	15.5	23.5	23.5	27.5	31.5	11.5
Shifts	S2	2→1	2→2	2→3	2→4	2→5	2→6	2→7
Rest hrs		11.5	11.5	19.5	19.5	23.5	27.5	7.5
Shifts	S3	3→1	3→2	3→3	3→4	3→5	3→6	3→7
Rest hrs		7.5	7.5	15.5	15.5	19.5	23.5	3.5
Shifts	S4	4→1	4→2	4→3	4→4	4→5	4→6	4→7
Rest hrs		3.5	3.5	11.5	11.5	15.5	19.5	-0.5
Shifts	S5	5→1	5→2	5→3	5→4	5→5	5→6	5→7
Rest hrs		-0.5	-0.5	7.5	7.5	11.5	15.5	-4.5
Shifts	S6	6→1	6→2	6→3	6→4	6→5	6→6	6→7
Rest hrs		-0.5	-0.5	7.5	7.5	11.5	15.5	-4.5
Shifts	S7	7→1	7→2	7→3	7→4	7→5	7→6	7→7
Rest hrs		15.5	15.5	23.5	23.5	27.5	31.5	11.5

Shifts: changes of shifts in two consecutive workdays, $i \rightarrow j$, the employee works in shift i of the first day and works shift j of the immediate following day; Rest hrs: Rest hours between consecutive shifts

The constraints (1), (2)–(6), (8), (9) are to fulfill the management objectives of the hospital, and the constraints (7), (10)–(12) are the requirements to satisfy the employees in the hospital.

3 Binary Integer Programming Formulation

3.1 Notations

3.1.1 Index

s : the shift index, $s \in \{1,2,3,4,5,6,7\}$ (Table 1).
t : the types of employees, $t \in \{1,2,3,4,5,6\}$ (Table 2).
w : the week index within 1 month, $w \in \{1,2,3,4\}$
d : the day within each week in the order of Monday, Tuesday, Wednesday, Thursday, Friday, Saturday, Sunday, $d = 1,2,3,4,5,6,7$, and the weekdays are the cases of $d = 1,2,3,4,5$, the weekends when $d = 6,7$
i_t : the ID of nurses within each type $t, i_t = 1,\ldots, Num_t$, where Num_t is defined in the following.

3.1.2 Hard Constraints Parameters

T_s : time length of shift s in hours, $T_1 = T_3 = T_6 = 8.5, T_2 = T_4 = T_5 = T_7 = 12.5$ (Table 1).

$n_s, N_s (nd_s, Nd_s)$: the min, max requirements of numbers of employees for the shift s within each weekday (weekend) (Table 4).

$n_{t,s}, N_{t,s} (nd_{t,s}, Nd_{t,s})$: the min, max requirement of numbers of employees for the shift s of type t within each weekday (weekend) (Table 3).

Num_t : the number of available employees within type $t, Num_1 = 29, Num_2 = 11, Num_3 + Num_4 = 7, Num_5 = Num_6 = 1$ (Table 2).

3.1.3 Soft Constraints Parameters

$Pref_{t,i_t,s,w}$: The preference of the employee i_t to work on the shift s of week w. This can be represented in a binary tensor with four dimensions (Table 5):

$$t \in \{1,\ldots,4\}, i_t \in \{1,\ldots,Num_g\}, s \in \{1,\ldots,7\}, w \in \{1,\ldots,4\}$$

Assume that Manager and Nurse Leader have no preference in this case.

$SevenDays_{t,i_t,w}$: The employee i_t prefers to work in the week w or not. This can be stored in a tensor of $\{0,1\}$ of three dimensions (Table 5):

$$t \in \{1,\cdots,4\}, i_t \in \{1,\cdots,Num_t\}, w \in \{1,\cdots,4\}$$

R_{t,i_t} : the priority rank matrix. The priority rank for the employee i_t of type t within all employees is used to decide who should be satisfied first if some of them have the same preferred shift within a week (Table 5).

C_{t,i_t} : the coefficient matrix in the object function. Usually, $C_{t,i_t} = a - bR_{t,i_t}$, where the coefficients a and b are chosen a priori. In our case, we have chosen $a = 1,500, b = 20$.

3.2 Decision Variables

$$x_{t,i_t,s,w,d} = \begin{cases} 0 \\ 1 \end{cases}$$

This is a binary variable to assign the employee i_t of type t to work in the shift s of day d in week w if 1; otherwise 0. This is an output tensor in our model with five dimensions:

$$\forall t \in \{1,\ldots,6\}, i_t \in \{1,\ldots,Num_t\}, s \in \{1,\ldots,7\}, w \in \{1,\ldots,4\}, d \in \{1,\ldots,7\}$$

3.3 Constraints

1. Daily shift requirement is that the number of employees for each shift should be satisfied (Table 4). The manager and clinical nurse leader are not counted for shifts.

Thus, for weekdays of shift s, we have

$$n_s \leq \sum_{t=1}^{4} \sum_{i_t=1}^{Num_t} x_{t,i_t,s,w,d} \leq N \qquad (1)$$

$$\forall s \in \{1,\ldots,7\}, w \in \{1,\ldots,4\}, d \in \{1,\ldots,5\}$$

and for weekends of shift s, we have

$$nd_s \leq \sum_{t=1}^{4} \sum_{i_t=1}^{Num_t} x_{t,i_t,s,w,d} \leq Nd \qquad (2)$$

$$\forall s \in \{1,\ldots,7\}, w \in \{1,\ldots,4\}, d \in \{6,7\}$$

Nurse Scheduling Problem: An Integer Programming Model with a Practical Application 73

In addition, there are the requirements of numbers of nurses within each type for different shifts (Table 3). These kinds of requirements are only for the shifts 1, 3, and 6.

Thus, for weekdays of shift s, we have

$$n_{t,s} \leq \sum_{i_t=1}^{Num_t} x_{t,i_t,s,w,d} \leq N_{t,} \tag{3}$$

$$\forall t \in \{1,\ldots,4\}, s \in \{1,3,6\}, w \in \{1,\ldots,4\}, d \in \{1,\ldots,5\}$$

and for weekends of shift s, we have

$$nd_{t,s} \leq \sum_{i_t=1}^{Num_t} x_{t,i_t,s,w,d} \leq Nd_{t,s} \tag{4}$$

$$\forall t \in \{1,\ldots,4\}, s \in \{1,3,6\}, w \in \{1,\ldots,4\}, d \in \{1,\ldots,5\}$$

2. RN $(t=1)$ may not work greater than 12 h per shift, and LPN, NA, HCT $(t=2,3,4)$ may not work greater than 16 h per shift. From Table 1, $T_1 = T_3 = T_6 = 8.5, T_2 = T_4 = T_5 = T_7 = 12.5$, no shift has the length larger than 12.5 h.

$$x_{1,i_1,s,w,d} = 0 \tag{5}$$

$$\forall i_1 \in \{1,\ldots,Num_1\}, s \in \{2,4,5,7\}, w \in \{1,\ldots,4\}, d \in \{1,\ldots,7\}$$

3. RN $(t=1)$ may not work greater than 60 h per week.

$$\sum_{d=1}^{7} \sum_{s=1}^{7} x_{1,i_1,s,w,d} T_s \leq 60 \tag{6}$$

$$\forall i_1 \in \{1,\ldots,Num_1\}, w \in \{1,\ldots,4\}$$

4. RN, LPN, NA, HCT $(t=1,2,3,4)$ may not work more than 6 days a week if he/she does not prefer to work 7 days in that week.

$$\sum_{d=1}^{7} \sum_{s=1}^{7} x_{t,i_t,s,w,d} \leq 6(1 - SevenDays_{t,i_t,w}) + 7 SevenDays_{t,i_t,w} \tag{7}$$

$$\forall t \in \{1,\ldots,4\}, i_t \in \{1,\ldots,Num_t\}, w \in \{1,\ldots,4\}$$

5. RN (t = 1) may not work more than 7 days consecutively. Considering every 8 days, an RN may work at most 7 days.

$$\sum_{d=i}^{7}\sum_{s=1}^{7} x_{1,i_1,s,w,d} + \sum_{d=1}^{i}\sum_{s=1}^{7} x_{1,i_1,s,w+1,d} \leq 7 \quad (8)$$

$$\forall i_1 \in \{1,\ldots,Num_1\}, w \in \{1,\ldots,3\}, i \in \{1,\ldots,7\}$$

6. All employees should work 4–6 weekend days per month (the nurse manager and clinical nurse leader do not need to do this).

$$4 \leq \sum_{w=1}^{4}\sum_{d=6}^{7}\sum_{s=1}^{7} x_{t,i_t,s,w,d} \leq 6 \quad (9)$$

$$\forall t \in \{1,\ldots,4\}, i_t \in \{1,\ldots,Num_g\}$$

7. The schedule of an employee should avoid shift overlap contradiction and also guarantee enough rest time between two consecutive shifts of that employee (Table 6);

As shown in Table 6, we list all possible of time off between any two shifts on the first day and its immediate following day, and also highlight that rest hours have lengths less than 10 h.

   7.1 The consecutive shifts, from shift 2 to shift 7 (2→7), shift 3 to shift 1 (3→1), shift 3 to shift 2 (3→2), shift 5 to shift 3 (5→3), shift 5 to shift 4 (5→4), shift 6 to 3 (6→3), and shift 6 to 4 (6→4), are all of length 7.5 h for rest;
   7.2 The consecutive shifts, shift 3 to shift 7 (3→7), shift 4 to shift 1 (4→1), and shift 4 to shift 2 (4→2), are all of length 3.5 h for rest;
   7.3 The consecutive shifts, shift 4 to shift 7 (4→7), shift 5 to shift 1 (5→1), shift 5 to shift 2 (5→2), shift 5 to shift 7 (5→7), shift 6 to shift 1 (6→1), shift 6 to shift 2 (6→2), and shift 6 to shift 7 (6→7), are all of negative lengths, which imply that the beginning time of the shift of next day is earlier before the ending time of the shift on the first day.

Since we assume that each employee works in the same shift of a week, we consider the changes of shift from Sunday to next Monday. Thus, these constraints are as follows.

$$x_{t,i_t,s_a,w-1,7} + x_{t,i_t,s_b,w,1} \leq 1 \quad (10)$$

$$\forall t \in \{1,\ldots,6\}, i_t \in \{1,\ldots,Num_g\}, w \in \{2,\ldots,4\}$$

$$(s_a, s_b) \in \{(2,7),(3,1),(3,2),(5,3),(5,4),(6,3),(6,4)\} \cup \{(3,7),(4,1),(4,2)\}$$
$$\cup \{(4,7),(5,1),(5,2),(5,7),(6,1),(6,2),(6,7)\}$$

8. The different time requirement for full-time and part-time employees (Table 2) should be satisfied. Assume that the first three nurses in the first type (RN) and the first nurse in the second type (LPN) work part time, and all other employees work full time. In the following, two cases are included: when $i = 1$, the requirements for weeks 1 and 2; when $i = 2$, the requirements for weeks 3 and 4. Here we define an upper bound for each time constraint since the sum of time lengths for shift is not exactly 80, 72, 48, and 24. Assume the upper bounds are 88, 80, 56, 32 for these cases, respectively. Thus, for $i \in \{1,2\}$, we have the full/part-time constraints:

$$80 \leq \sum_{w=i}^{i+1} \sum_{d=1}^{7} \sum_{s=1}^{7} x_{t,i_t,s,w,d} T_s \leq 88 \quad (11)$$

$$\forall t \in \{1,\ldots,6\}, i_1 \in \{4,\ldots,Num_1\}, i_2 \in \{2,\ldots,Num_2\},$$

$$i_t \in \{1,\ldots,Num_t\} \text{ for } t = 3,4,5,6$$

$$72 \leq \sum_{w=i}^{i+1} \sum_{d=1}^{7} \sum_{s=1}^{7} x_{1,1,s,w,d} T_s \leq 80 \quad (12)$$

$$48 \leq \sum_{w=i}^{i+1} \sum_{d=1}^{7} \sum_{s=1}^{7} x_{1,1,s,w,d} T_s \leq 56 \quad (13)$$

$$t = 1, i_1 = 2 \text{ and } t = 2, i_2 = 1$$

$$24 \leq \sum_{w=i}^{i+1} \sum_{d=1}^{7} \sum_{s=1}^{7} x_{1,3,s,w,d} T_s \leq 32 \quad (14)$$

9. The manager and the clinical nurse leader $(t = 5,6)$ present day shift (assume $s = 1$ is the only day shift) only.

$$x_{t,i_t,s,w,d} = 0 \quad (15)$$

$$\forall t \in \{5,6\}, i_5 = i_6 = 1, s \in \{2,\ldots,7\}, w \in \{1,\ldots,4\}, d \in \{1,\ldots,7\}$$

10. Each employee works at most one shift per day.

$$\sum_{s=1}^{7} x_{t,i_t,s,w,d} \leq 1 \quad (16)$$

$$\forall t \in \{1,\ldots,6\}, i_t \in \{1,\ldots,Num_g\}, w \in \{1,\ldots,4\}, d \in \{1,\ldots,7\}$$

11. Each employee (t, i_t) each week w works only on one shift s. Defining another binary variable $y_{t,i_t,s,w} = 0,1$ to indicate the employee i_t of type t is working in shift s of week w or not. Thus, $y_{t,i_t,s,w}$ and $x_{t,i_t,s,w,d}$ have the relations as follows:

$$y_{t,i_t,s,w} \geq x_{t,i_t,s,w,d} \tag{17}$$

$$\forall t \in \{1,\ldots,6\}, i_t \in \{1,\ldots,Num_g\}, s \in \{1,\ldots,7\}, w \in \{1,\ldots,4\}, d \in \{1,\ldots,7\}$$

$$\sum_{s=1}^{7} y_{t,i_t,s,w} \leq 1 \tag{18}$$

$$\forall t \in \{1,\ldots,6\}, i_t \in \{1,\ldots,Num_g\}, w \in \{1,\ldots,4\}$$

12. Some employees may need vacation time. In this model, we assume that two employees have the requirements of time off: the 28th nurse of RN takes off on week 2 and week 3; and 7th nurse of LPN takes off on week 3 and 4.

$$x_{1,28,s,w,d} = 0 \tag{19}$$

$$\forall s \in \{1,\ldots,7\}, w \in \{2,3\}, d \in \{1,\ldots,7\}$$

and this may influence the constraints (8). We assume the total working hours is based on the week 1 and week 4. For the time off of the 7th nurse of type LPN, we have the constraints

$$x_{2,28,s,w,d} = 0 \tag{20}$$

$$\forall s \in \{1,\ldots,7\}, w \in \{3,4\}, d \in \{1,\ldots,7\}$$

and this also influences the constraints (8). We assume the total working hours is based on week 1 and week 2.

3.4 Objective Functions

The objective of NSP is to maximize the satisfaction of all employees. The manager and the nurse leader are not considered in the objective function and the constraints will ensure to present a feasible schedule for both of them. We consider the satisfactions based on the previously chosen preferences of all employees as following two cases.

A. In case we have one preference shift for all days in the week, the objective function is the following:

$$\max \sum_{t=1}^{4} \sum_{i_t=1}^{Num_t} \sum_{s=1}^{7} \sum_{w=1}^{4} C_{t,i_t} Pref_{t,i_t,s,w} y_{t,i_t,s,w} \qquad (21)$$

where C_{t,i_t} is an coefficient related to the priority rank of employee i_t of type t (Table 5) and $Pref_{t,i_t,s,w} y_{t,i_t,s,w}$ is 1 if the preferred shift s of week w for employee i_t of type t is 1 if he/she is satisfied and 0 otherwise.

B. In case we have two preference shifts in a week (one for weekdays and one for weekends), the objective function is the following:

$$\max \sum_{t=1}^{4} \sum_{i_t=1}^{Num_t} \sum_{s=1}^{7} \sum_{w=1}^{4} \sum_{c=1}^{2} C_{t,i_t} Pref_{t,i_t,s,w,c} y_{t,i_t,s,w,c}, \qquad (22)$$

where $Pref_{t,i_t,s,w,c}$ is the preference of the employee i_t to work on the shift s of week w on weekday (if $c = 1$) or on weekend (if $c = 2$); $y_{t,i_t,s,w,c}$ and $x_{t,i_t,s,w,d,c}$ are additional variables, similar to $y_{t,i_t,s,w}$ and $x_{t,i_t,s,w,d}$, respectively, taking into account the division of the week to weekdays and weekend, and satisfying analogous requirements (see (17)). Similar analysis can be changed on constraints (see (10)) from Friday and Saturday.

The optimization model for NSP is based on the programming with (21), (22) as objective function and (1)–(20) as constraints. This is a binary linear integer program.

4 Implementations and a Practical Example

The algorithm for optimization model with objective (21) or (22) and constraints (1)–(20) are implemented using CPLEX 11.0 (2008) via ILOG Concert Technology 2.5, and all computations are performed on a SUN UltraSpace-III with a 900 MHz processor and 2.0 GB RAM. In the following, we present part results. Since the total number of NA and HCT is 7, we split 4 for NA and 3 for HCT.

We implemented two types of preferences. The first defines a desired shift every day in a week (see Appendix 1). The complete schedule with respect to this kind of preference is in Appendix 2.

The second defines a desired shift for the weekdays (Monday to Friday) and another desired shift for the weekend (see Appendix 3). The second type of preferences is more flexible and gives more satisfaction. The complete schedule with respect to this kind of preference is in Appendix 4.

In the following, we present part of our results in tables. Table 7 is an example of input preferences for employees (only first 5 nurses of RN, LPN are listed); Table 8 is a schedule for employees; Table 9 is the total working hours per two weeks; and Figure 1 and Figure 2 presents the satisfaction of the schedule compared with the employees' preferences. For complete results, we present in Appendix 1 (preferences) and Appendix 2 (schedule).

Table 7 An example of input preferences for employees from RN and LPN

Type	Member ID	Week 1	Week 2	Week 3	Week 4
RN	1	S6	S6	S3	S1
	2	S1	S1	S6	S6
	3	S6	S3	S3	S3
	4	S1	S3	S6	S3
	5	S1	S6	S3	S1
LPN	1	S4	S4	S2	S3
	2	S5	S5	S2	S5
	3	S4	S6	S2	S2
	4	S1	S5	S2	S7
	5	S6	S3	S3	S5
NA	1	S7	S5	S2	S2
	2	S7	S5	S5	S5
	3	S2	S7	S2	S5
	4	S2	S5	S7	S5
HCT	1	S7	S2	S5	S2
	2	S7	S2	S2	S2
	3	S5	S4	S3	S5

In Table 7, we present the preferred shift for each employee on every week (For RN and LPN, only first five nurses and for others, check Appendix 1). The empty cell means there is no preference for the corresponding employee for that week. Next, we present a possible schedule for these nurses in Table 8 by our model solved via CPLEX.

In Table 8, each cell represents the shift that corresponding nurse works on that specific day, and an empty cell means that nurse has rest on the corresponding day. As we have mentioned in constraints (8), the first three nurses in RN and the first in LPN are part-time nurses and all others are full-time. In Table 9, we present the results of their working hours obtained by our model. For other nurses of types RN and LPN, we list in Appendix 2. The type NM denotes the nurse manager and NL denotes the nurse leader.

If the scheduled shift obtained by the model is the same as the employee's choice before scheduling, this is satisfied for the employee; otherwise, not. We compute the percentage of satisfactions (total number of satisfied assignment over all assignment) for different types in Figures 1 and 2.

5 Conclusion

We developed a new nurse scheduling problem aiming to satisfy specific schedule requirements in a VA hospital and maximally satisfy the preferences of involved nurses and other type of employees. Besides the common requirements for NSP,

Nurse Scheduling Problem: An Integer Programming Model with a Practical Application

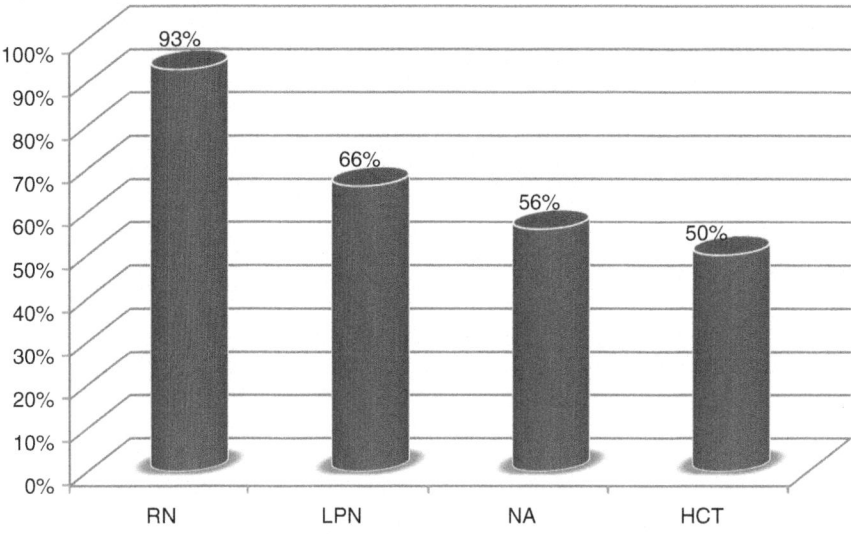

Fig. 1 The percentage of satisfactions for types (one preference for a whole week)

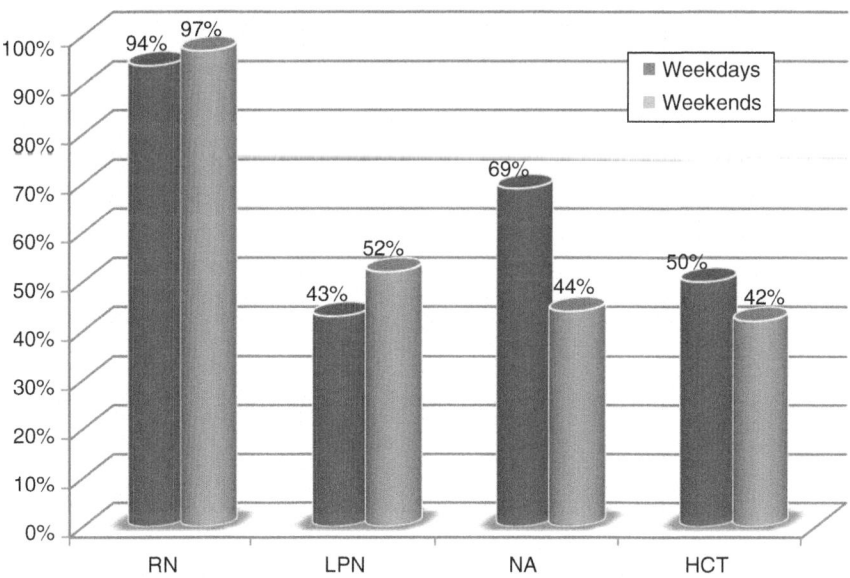

Fig. 2 The average percentage of satisfactions for types with two preference shift (one for weekdays and one for weekends)

Table 8 A schedule for employees (one preference for a whole week)

	ID	Week 1							Week 2							Week 3							Week 4						
		M	T	W	T	F	S	S	M	T	W	T	F	S	S	M	T	W	T	F	S	S	M	T	W	T	F	S	S
RN	1	6	6				6	6	6	6	6	6	6	6		3	3	3	3	3	3		1						
	2		1	1	1			1						1	1		6	6	6	6	6	6					6		1
	3							6		3	3		3	3	3	3	3		3		3	3	3	3	3	3	3	3	3
	4	1	1	1	1	1	1		3	3		6	4	4													1	1	1
	5	1	1						6	6	4	5	5		5							6	5	5	5	5	5	5	5
LPN	1									5		5	5	4			2		2	2	4	4	5	5	5	5	5	5	5
	2	4		5	5	5	4	5	5	3	5	3	5	3		2	2	2	2	2	2		2	2	2	2	2	2	2
	3	4	4	4	4	4	4	4			5		3		6				2										
	4						7		5	5	5	5	5	5	3			2		2	2	3	5	5	5	5	5	5	5
	5	6		6	6	7	7		3	3	3	5	3	3	5			2	2	2		5	2	2		2	2		2
NA	1			7	7				5	5	5	5	5	5		5	5	5	5	5	5	5	5	5	5	5	5	5	5
	2						2		5	5						2		2		2	2	2	2	2		2	2	2	2
	3		2	2	2	2	2		7	7					5	7	7	7	5	2	2		5	5		5	5		5
	4	2	2	2	2	2					2		2	2	2	2	5	5	7	7	7	7							
HCT	1	7	7	5			5		2	2	2		2			5	5	5	5	5	5	5							
	2							7	2	2	2	2		2	2	2	2		2	2	2		2		2		2	2	2
	3	5	5	5					4			4					3	3	3	3	3								5
NM		1	1	1	1	1	1		1	1	1	1	1	1	1	1	1	1	1	1	1	1	1	1					1
NL		1	1	1	1	1	1									1	1	1	1	1	1	1	1	1			1	1	1

Table 9 Working hours per 2 weeks (in hours)

Type	Member ID	Required hours per 2 weeks	Working hours	
			Week 1 and week 2	Week 3 and week 4
RN	1	72	76.5	76.5
	2	48	51	51
	3	24	25.5	25.5
	4	80	85	85
	5	80	85	85
LPN	1	48	50	50
	2	80	87.5	87.5
	3	80	83.5	87.5
	4	80	83.5	87.5
	5	80	85	87.5
NA	1	80	87.5	87.5
	2	80	87.5	87.5
	3	80	87.5	87.5
	4	80	87.5	87.5
HCT	1	80	87.5	87.5
	2	80	87.5	87.5
	3	80	87.5	80.5
Nurse manager		80	85	85
Nurse leader		80	85	85

we consider specific constraints in our model imposed by the VA hospital, such as the number required by each shift with respect to different type of nurses, the longest time that a nurse can work on a shift, the possibilities of a nurse to work 7 days in a week if she/he wants, the weekends requirement, full-time or part-time requirements for different employees, and so on. In addition, we consider two supervisors, the nurse manager and the clinical nurse leader in our model whose requirements are different, such as they present only on day shifts. The holidays or emergency leave of nurses are also considered as constraints. We implemented two types of preferences: the first defines a desired shift every day in a week, and the second defines a desired shift for the weekdays (Monday to Friday) and another desired shift for the weekend. The second type of preferences is more flexible and gives more satisfaction. Although our model is based on a period of a month consisting of 4 weeks, it can be easily extended to a long period such as a year with 52 weeks. Another direction for further research we can concentrate on is to find the minimum number of nurses to satisfy all the requirements for all shifts while also maximizing the nurse satisfaction.

Acknowledgment This material is based upon work supported by the Office of Systems Redesign, Department of Veteran Affairs. This research has been partially supported by CAO funds.

6 Appendix 1: Results of Nurse Scheduling with One Preference Shift in a Week

	Preferences				Assignments				Satisfaction			
Weeks	1	2	3	4	1	2	3	4	1	2	3	4
Type 1 (RN)												
1	6	6	3	1	6	6	3	1	Y	Y	Y	Y
2	1	1	6	6	1	1	6	6	Y	Y	Y	Y
3	6	3	3	3	6	3	3	1	Y	Y	Y	N
4	1	3	6	3	1	3	6	3	Y	Y	Y	Y
5	1	6	3	1	1	6	1	1	Y	Y	N	Y
6	6	1	1	6	6	1	1	6	Y	Y	Y	Y
7	6	6	1	1	6	6	1	1	Y	Y	Y	Y
8	3	1	6	6	3	1	6	6	Y	Y	Y	Y
9	1	6	1	6	1	6	1	6	Y	Y	Y	Y
10	6	1	6	3	6	1	6	3	Y	Y	Y	Y
11	3	3	6	1	3	3	6	1	Y	Y	Y	Y
12	3	3	6	3	3	3	6	3	Y	Y	Y	Y
13	6	6	6	3	6	6	6	3	Y	Y	Y	Y
14	3	3	1	6	3	3	1	6	Y	Y	Y	Y
15	6	6	1	1	6	6	1	1	Y	Y	Y	Y
16	3	1	1	6	3	1	1	6	Y	Y	Y	Y
17	3	1	1	1	3	1	1	1	Y	Y	Y	Y
18	3	3	1	1	3	3	1	1	Y	Y	Y	Y
19	1	1	6	6	1	1	6	6	Y	Y	Y	Y
20	1	6	1	6	1	6	1	6	Y	Y	Y	Y
21	6	3	6	6	6	3	6	6	Y	Y	Y	Y
22	3	3	3	1	3	3	3	1	Y	Y	Y	Y
23	6	1	6	3	6	1	6	3	Y	Y	Y	Y
24	6	3	3	1	6	3	1	1	Y	Y	N	Y
25	3	1	3	3	3	1	6	3	Y	Y	N	Y
26	3	1	3	1	3	1	1	1	Y	Y	N	Y
27	1	6	3	3	1	6	3	3	Y	Y	Y	Y
28	3	1	1	6	3	3	3	6	Y	N	N	Y
29	1	1	3	3	1	1	6	3	Y	Y	N	Y

(continued)

Appendix 1 (continued)

	Preferences				Assignments				Satisfaction			
Weeks	1	2	3	4	1	2	3	4	1	2	3	4
Type 2 (LPN)												
1	4	4	2	3	1	4	2	3	N	Y	Y	Y
2	5	5	2	5	5	5	4	5	Y	Y	N	Y
3	4	6	2	2	4	6	1	6	Y	Y	N	N
4	1	5	2	7	1	5	2	7	Y	Y	Y	Y
5	6	3	3	5	6	3	3	5	Y	Y	Y	Y
6	5	4	2	4	5	6	1	4	Y	N	N	Y
7	6	7	1	1	6	7	4	5	Y	Y	N	N
8	4	2	4	5	4	1	4	3	Y	N	Y	N
9	2	2	5	2	6	1	5	4	N	N	Y	N
10	5	5	6	7	5	5	6	7	Y	Y	Y	Y
11	2	7	4	2	2	6	4	6	Y	N	Y	N
Type 3 (NA)												
1	7	5	2	2	7	5	1	7	Y	Y	N	N
2	7	5	5	5	1	5	5	6	N	Y	Y	N
3	2	7	2	5	2	7	4	6	Y	Y	N	N
4	2	5	7	5	2	5	7	7	Y	Y	Y	N
Type 4 (HCT)												
1	7	2	5	2	3	3	5	7	N	N	Y	N
2	7	2	2	2	7	2	6	1	Y	Y	N	N
3	5	4	3	5	5	4	3	6	Y	Y	Y	N

7 Appendix 2: Results of Assignments with Respect to Each Day with One Preference Shift in a Week (Employees Who Do Not Appear in this Table Mean They Do Not Need to Work in that Corresponding Week)

Week 1							
(ID.Shift)	Monday	Tuesday	Wednesday	Thursday	Friday	Saturday	Sunday
Type 1 RN							
1.6	1	1					1
2.1	1	1	1		1	1	1
3.6							1
4.1	1	1	1	1		1	1
5.1	1		1	1	1	1	1
6.6	1	1	1		1		
7.6			1	1	1		1
8.3		1	1	1	1		
9.1	1	1	1	1	1		1
10.6	1	1			1	1	
11.3	1			1	1	1	
12.3	1		1		1		1
13.6	1		1	1		1	1
14.3		1	1	1			1
15.6		1	1	1			1
16.3	1	1		1	1	1	
17.3	1	1		1	1		
18.3	1	1	1		1	1	1
19.1	1	1	1	1	1	1	
20.1	1	1		1	1	1	1
21.6	1	1	1			1	
22.3	1	1				1	1
23.6	1	1		1	1	1	
24.6		1	1	1	1	1	
25.3			1	1	1	1	1
26.3	1		1		1	1	
27.1	1	1	1	1	1	1	
28.3	1	1	1	1		1	1
29.1		1	1	1	1	1	1
Type 2 (LPN)							
2.5				1	1	1	1
3.4	1	1		1	1	1	1
4.1							1
5.6	1		1	1			
6.5	1	1	1	1	1	1	1
7.6					1	1	1
8.4			1				

(continued)

Appendix 2 (continued)

Week 1							
(ID.Shift)	Monday	Tuesday	Wednesday	Thursday	Friday	Saturday	Sunday
9.2	1					1	1
10.5	1	1	1	1	1		1
11.2	1	1	1	1	1	1	1
Type 3 (NA)							
1.7			1	1	1	1	
3.2		1	1	1	1	1	
4.2	1	1	1	1	1		1
Type 4 (HCT)							
1.7	1	1					
2.7							1
3.5	1	1	1			1	
Week 2							
	Monday	Tuesday	Wednesday	Thursday	Friday	Saturday	Sunday
Type 1 (RN)							
1.6	1	1	1	1	1	1	
3.3						1	1
4.3		1	1	1		1	
5.6	1	1		1	1		
6.1	1		1	1	1	1	1
7.6	1	1	1	1		1	1
8.1	1	1	1	1	1	1	
9.6			1	1	1	1	
10.1	1	1	1		1	1	1
11.3	1	1		1	1	1	1
12.3	1	1		1	1	1	1
13.6	1	1	1		1		1
14.3	1	1	1	1	1	1	
15.6	1	1	1	1	1	1	
16.1	1		1	1	1		1
17.1	1	1	1		1	1	1
18.3	1		1		1	1	
19.1	1		1	1	1		
20.6		1	1	1			1
21.3	1	1	1	1	1		1
22.3	1	1	1	1	1		1
23.1	1	1		1		1	1
24.3		1	1	1	1		1
25.1		1		1	1	1	1
26.1	1	1	1	1		1	1
27.6	1				1	1	1
29.1		1		1		1	1
Type 2 (LPN)							
1.4			1		1	1	1
2.5	1			1	1		
3.6							1

(continued)

Appendix 2 (continued)

Week 1							
(ID.Shift)	Monday	Tuesday	Wednesday	Thursday	Friday	Saturday	Sunday
4.5	1	1	1	1	1	1	
5.3	1	1	1	1	1	1	1
7.7			1	1	1	1	1
8.2	1	1		1	1	1	1
9.2	1		1	1	1		
10.5							1
Type 3 (NA)							
1.5		1	1				1
2.5	1	1	1	1	1	1	1
3.7	1	1					
4.5						1	
Type 4 (HCT)							
1.2		1	1		1	1	1
2.2	1	1	1	1		1	1
3.4	1	1		1			
Week 3							
	Monday	Tuesday	Wednesday	Thursday	Friday	Saturday	Sunday
Type 1 (RN)							
1.3	1	1	1	1	1	1	
2.6		1	1		1	1	1
3.3		1				1	1
4.6	1		1		1	1	
5.3	1		1	1		1	
6.1	1	1		1	1	1	1
7.1		1	1	1		1	1
8.6	1			1		1	1
9.1	1		1	1	1	1	1
10.6		1		1	1	1	
11.6		1		1	1		1
12.6	1		1		1	1	
13.6	1	1	1	1			
14.1	1	1	1		1	1	1
15.1	1	1		1	1		
16.1	1	1	1	1	1	1	
17.1	1		1	1		1	1
18.1	1	1	1				1
19.6	1	1	1		1	1	1
20.1		1	1	1	1	1	1
21.6	1		1	1			1
22.3	1	1	1	1	1	1	
23.6	1	1		1	1		
24.3	1		1	1	1		1
25.3	1	1			1		1
26.3	1	1	1	1	1		1

Appendix 2 (continued)

Week 1							
(ID.Shift)	Monday	Tuesday	Wednesday	Thursday	Friday	Saturday	Sunday
27.3		1	1	1	1	1	
29.3	1		1		1		1
Type 2 (LPN)							
1.2		1	1			1	1
3.2				1			
4.2	1	1	1	1	1	1	1
8.4	1		1	1	1	1	1
9.5	1	1	1	1	1	1	1
11.4		1					
Type 3 (NA)							
1.2			1				
2.5	1	1	1	1	1	1	1
3.2	1				1	1	1
4.7	1	1	1	1	1	1	1
Type 4 (HCT)							
1.5	1	1	1	1	1	1	1
2.2	1	1		1	1		
3.3		1	1	1		1	1
Week 4							
	Monday	Tuesday	Wednesday	Thursday	Friday	Saturday	Sunday
Type 1 (RN)							
1.1	1					1	1
2.6					1		
4.3	1	1	1		1	1	1
5.1	1	1	1		1	1	1
6.6	1		1			1	1
7.1	1		1	1	1		1
8.6	1	1	1	1	1		1
9.6	1	1		1	1		
10.3	1	1	1	1	1	1	
11.1		1	1	1	1	1	1
12.3	1	1		1	1	1	
13.3	1	1	1	1	1		1
14.6	1	1		1		1	
15.1	1	1		1	1	1	1
16.6	1		1	1			1
17.1	1	1		1		1	
18.1	1	1	1	1	1	1	
19.6		1	1	1		1	
20.6	1		1		1	1	
21.6	1	1		1	1		1
22.1	1		1		1		1
23.3		1	1	1	1	1	1
24.1		1	1	1	1		1
25.3	1	1	1	1	1	1	

(continued)

Appendix 2 (continued)

Week 1							
(ID.Shift)	Monday	Tuesday	Wednesday	Thursday	Friday	Saturday	Sunday
26.1		1	1	1		1	
27.3	1		1	1		1	1
28.6		1			1	1	1
29.3	1	1	1	1	1		1
Type 2 (LPN)							
2.5	1	1	1	1	1	1	1
3.2	1	1	1	1	1	1	
5.5	1	1	1	1	1	1	1
6.4	1	1	1	1	1	1	1
8.5	1						
10.7	1	1	1	1	1	1	1
11.2	1	1	1	1		1	1
TYPE 3 (NA)							
1.2	1	1		1	1	1	1
3.5		1	1			1	
Type 4 (HCT)							
2.2			1		1		1
3.5				1	1		1

8 Appendix 3: Results of Nurse Scheduling with Two Preferences Shift (One for Weekdays and One for Weekends)

Weeks	Preferences				Assignments				Satisfaction			
	1	2	3	4	1	2	3	4	1	2	3	4
Type 1 (RN)												
1	(6,6)	(6,1)	(3,6)	(1,3)	(6,6)	(6,1)	(3,6)	(1,3)	(Y,Y)	(Y,Y)	(Y,Y)	(Y,Y)
2	(1,6)	(1,3)	(6,1)	(6,6)	(1,1)	(1,3)	(6,1)	(6,6)	(Y,N)	(Y,Y)	(Y,Y)	(Y,Y)
3	(6,6)	(3,1)	(3,6)	(1,6)	(3,6)	(3,1)	(3,6)	(1,6)	(N,Y)	(Y,Y)	(Y,Y)	(Y,Y)
4	(1,3)	(3,3)	(6,6)	(3,1)	(1,3)	(3,3)	(6,6)	(3,1)	(Y,Y)	(Y,Y)	(Y,Y)	(Y,Y)
5	(1,6)	(6,6)	(1,6)	(1,3)	(1,6)	(6,6)	(1,6)	(1,3)	(Y,Y)	(Y,Y)	(Y,Y)	(Y,Y)
6	(6,1)	(1,3)	(1,1)	(6,3)	(6,1)	(1,3)	(1,1)	(6,3)	(Y,Y)	(Y,Y)	(Y,Y)	(Y,Y)
7	(6,6)	(6,3)	(1,1)	(1,3)	(6,6)	(6,3)	(1,1)	(1,3)	(Y,Y)	(Y,Y)	(Y,Y)	(Y,Y)
8	(3,6)	(1,3)	(6,6)	(6,1)	(3,6)	(1,3)	(6,6)	(6,1)	(Y,Y)	(Y,Y)	(Y,Y)	(Y,Y)
9	(1,6)	(6,1)	(1,6)	(6,6)	(1,6)	(6,1)	(1,6)	(6,6)	(Y,Y)	(Y,Y)	(Y,Y)	(Y,Y)
10	(6,1)	(1,3)	(6,3)	(3,3)	(6,1)	(1,3)	(6,3)	(3,3)	(Y,Y)	(Y,Y)	(Y,Y)	(Y,Y)
11	(3,3)	(3,1)	(6,6)	(1,1)	(3,3)	(3,1)	(6,6)	(1,1)	(Y,Y)	(Y,Y)	(Y,Y)	(Y,Y)
12	(3,3)	(3,1)	(6,3)	(3,6)	(3,3)	(3,1)	(6,3)	(3,6)	(Y,Y)	(Y,Y)	(Y,Y)	(Y,Y)
13	(6,3)	(6,1)	(6,3)	(3,3)	(6,3)	(6,1)	(6,3)	(3,3)	(Y,Y)	(Y,Y)	(Y,Y)	(Y,Y)
14	(3,1)	(3,3)	(1,1)	(6,1)	(3,1)	(3,3)	(1,1)	(6,1)	(Y,Y)	(Y,Y)	(Y,Y)	(Y,Y)
15	(6,6)	(6,6)	(1,6)	(1,6)	(6,6)	(6,6)	(1,6)	(1,6)	(Y,Y)	(Y,Y)	(Y,Y)	(Y,Y)
16	(3,1)	(1,1)	(1,3)	(6,1)	(3,1)	(1,1)	(1,3)	(6,1)	(Y,Y)	(Y,Y)	(Y,Y)	(Y,Y)
17	(3,1)	(1,3)	(1,6)	(1,6)	(3,1)	(1,3)	(1,6)	(1,6)	(Y,Y)	(Y,Y)	(Y,Y)	(Y,Y)
18	(3,3)	(3,1)	(1,1)	(1,6)	(3,3)	(3,1)	(1,1)	(1,6)	(Y,Y)	(Y,Y)	(Y,Y)	(Y,Y)
19	(1,3)	(1,1)	(6,3)	(6,1)	(1,3)	(1,1)	(6,3)	(6,1)	(Y,Y)	(Y,Y)	(Y,Y)	(Y,Y)
20	(1,1)	(6,3)	(1,1)	(6,3)	(1,1)	(6,3)	(1,1)	(6,3)	(Y,Y)	(Y,Y)	(Y,Y)	(Y,Y)
21	(6,1)	(3,1)	(6,6)	(6,6)	(6,1)	(3,1)	(6,6)	(6,6)	(Y,Y)	(Y,Y)	(Y,Y)	(Y,Y)
22	(3,3)	(3,6)	(3,1)	(1,3)	(3,3)	(3,6)	(3,1)	(1,3)	(Y,Y)	(Y,Y)	(Y,Y)	(Y,Y)
23	(6,1)	(1,3)	(6,6)	(3,3)	(6,1)	(1,3)	(6,6)	(3,3)	(Y,Y)	(Y,Y)	(Y,Y)	(Y,Y)
24	(6,1)	(3,3)	(1,6)	(1,1)	(6,1)	(3,1)	(3,6)	(1,1)	(Y,Y)	(Y,N)	(N,Y)	(Y,Y)
25	(3,1)	(1,6)	(6,3)	(3,1)	(3,1)	(1,6)	(3,3)	(3,1)	(Y,Y)	(Y,Y)	(N,Y)	(Y,Y)
26	(3,6)	(1,3)	(1,1)	(1,6)	(3,6)	(1,3)	(3,1)	(1,6)	(Y,Y)	(Y,Y)	(N,Y)	(Y,Y)
27	(1,3)	(6,1)	(3,3)	(3,1)	(1,3)	(6,1)	(3,3)	(3,1)	(Y,Y)	(Y,Y)	(Y,Y)	(Y,Y)
28	(3,6)	(3,6)	(3,1)	(6,6)	(3,6)	(1,1)	(1,1)	(6,6)	(Y,Y)	(N,N)	(N,Y)	(Y,Y)
29	(1,3)	(1,1)	(6,1)	(3,3)	(1,3)	(1,1)	(3,1)	(3,3)	(Y,Y)	(Y,Y)	(N,Y)	(Y,Y)
Type 2 (LPN)												
1	(1,6)	(4,6)	(2,4)	(3,4)	(2,6)	(7,6)	(2,4)	(2,2)	(N,Y)	(N,Y)	(Y,Y)	(N,N)
2	(5,3)	(5,6)	(4,4)	(5,5)	(5,4)	(2,6)	(7,1)	(5,5)	(Y,N)	(N,Y)	(N,N)	(Y,Y)
3	(4,6)	(6,5)	(1,5)	(6,7)	(4,1)	(4,5)	(5,5)	(5,5)	(Y,N)	(N,Y)	(N,Y)	(N,N)
4	(1,7)	(5,3)	(2,1)	(7,7)	(2,7)	(5,2)	(2,5)	(2,7)	(N,Y)	(Y,N)	(Y,N)	(N,Y)
5	(6,3)	(3,4)	(3,5)	(5,6)	(5,3)	(3,4)	(3,1)	(5,1)	(N,Y)	(Y,Y)	(Y,N)	(Y,N)
6	(5,6)	(6,3)	(1,5)	(4,4)	(5,1)	(6,3)	(5,5)	(4,4)	(Y,N)	(Y,Y)	(N,Y)	(Y,Y)
7	(6,5)	(7,7)	(4,4)	(5,6)	(4,5)	(7,7)	(1,1)	(1,1)	(N,Y)	(Y,Y)	(N,N)	(N,N)
8	(4,6)	(1,4)	(4,2)	(3,6)	(7,4)	(2,2)	(4,2)	(4,2)	(N,N)	(N,N)	(Y,Y)	(N,N)
9	(6,2)	(1,7)	(5,2)	(4,6)	(2,2)	(2,2)	(5,2)	(5,5)	(N,Y)	(N,N)	(Y,Y)	(N,N)
10	(5,4)	(5,4)	(6,2)	(7,6)	(5,4)	(5,6)	(2,2)	(7,2)	(Y,Y)	(Y,N)	(N,Y)	(Y,N)
11	(2,7)	(6,2)	(4,7)	(6,3)	(2,1)	(4,2)	(2,7)	(2,2)	(Y,N)	(N,Y)	(N,Y)	(N,N)

(continued)

Appendix 3 (continued)

	Preferences				Assignments				Satisfaction			
Weeks	1	2	3	4	1	2	3	4	1	2	3	4
Type 3 (NA)												
1	(7,3)	(5,3)	(1,6)	(7,4)	(7,5)	(5,3)	(2,2)	(7,2)	(Y,N)	(Y,Y)	**(N,N)**	(Y,N)
2	(1,1)	(5,5)	(5,3)	(6,5)	(4,1)	(5,5)	(5,2)	(2,5)	(N,Y)	(Y,Y)	(Y,N)	(N,Y)
3	(2,4)	(7,5)	(4,2)	(6,3)	(2,2)	(7,5)	(4,2)	(2,5)	(Y,N)	(Y,Y)	(Y,Y)	**(N,N)**
4	(2,3)	(5,2)	(7,1)	(7,1)	(2,5)	(5,2)	(7,4)	(2,2)	(Y,N)	(Y,Y)	(Y,N)	**(N,N)**
Type 4 (HCT)												
1	(3,5)	(3,4)	(5,1)	(7,4)	(2,5)	(2,2)	(5,5)	(7,5)	(N,Y)	**(N,N)**	(Y,N)	(Y,N)
2	(7,2)	(2,6)	(6,5)	(1,1)	(5,2)	(2,6)	(2,5)	(2,5)	(N,Y)	(Y,Y)	(N,Y)	**(N,N)**
3	(5,2)	(4,7)	(3,4)	(6,3)	(5,2)	(4,1)	(3,5)	(4,4)	(Y,Y)	(Y,N)	(Y,N)	**(N,N)**

9 Appendix 4: Results of Assignments with Respect to Each Day with Two Preferences for Weekdays and Weekends (Employees Who Do Not Appear in this Table Mean They Do Not Need to Work in that Corresponding Week)

Week 1							
(ID.Shift)	Monday	Tuesday	Wednesday	Thursday	Friday	Saturday	Sunday
Type 1 (RN)							
1.6	1			1		1	
2.1	1		1	1	1	1	
3.6							1
4.1	1	1	1	1	1		
4.3							1
5.1	1	1	1	1	1		
5.6							1
6.1						1	1
6.6		1	1	1			
7.6	1	1		1	1	1	
8.3			1	1	1		
8.6						1	
9.1	1	1	1	1			
10.1							1
10.6			1	1	1		
11.3	1		1	1	1	1	
12.3	1	1				1	1
13.3							1
13.6	1	1	1		1		
14.1						1	1
14.3	1	1	1				
15.6	1	1	1	1	1		1
16.3	1	1	1	1	1		
17.1						1	1
17.3	1		1	1			
18.3			1	1	1	1	
19.1	1	1	1	1	1		
19.3						1	
20.1	1	1	1	1	1	1	
21.1						1	1
21.6	1		1				
22.3	1	1		1	1	1	1
23.1						1	1
23.6	1	1	1	1			
24.1							1
24.6		1	1	1	1		
25.3	1	1	1	1			
26.3		1	1		1		

(continued)

Appendix 4 (continued)

Week 1							
(ID.Shift)	Monday	Tuesday	Wednesday	Thursday	Friday	Saturday	Sunday
26.6						1	1
27.1		1	1	1	1		
27.3						1	1
28.3	1	1		1	1		
28.6						1	1
29.1	1	1		1	1		
29.3							1
Type 2 (LPN)							
1.2	1	1					
1.6							1
2.4							1
2.5					1		
3.4	1	1	1		1		
4.7						1	1
5.3						1	1
5.5	1		1	1			
6.1							1
6.5	1	1		1	1		
7.4				1			
7.5						1	1
8.4						1	
8.7	1	1					
9.2					1	1	1
10.5		1	1		1		
11.2	1	1		1	1		
Type 3 (NA)							
1.5							1
1.7			1	1	1		
3.2			1			1	
4.2	1	1	1	1	1		
4.5						1	
Type 4 (HCT)							
1.2			1	1			
1.5						1	1
2.2							1
2.5			1	1			
3.2						1	1
3.5	1	1					
Week 2							
	Monday	Tuesday	Wednesday	Thursday	Friday	Saturday	Sunday
Type 1 (RN)							
1.1							1
1.6	1	1	1	1	1		
2.3							1

(continued)

Appendix 4 (continued)

Week 1							
(ID.Shift)	Monday	Tuesday	Wednesday	Thursday	Friday	Saturday	Sunday
3.1						1	1
4.3			1	1	1	1	
5.6		1		1	1		1
6.1	1		1	1	1		
6.3						1	
7.3							1
7.6		1	1	1	1		
8.1	1	1	1	1	1		
8.3							1
9.1						1	1
9.6	1	1	1	1			
10.1	1	1	1	1	1		
10.3						1	
11.3	1	1	1	1	1		
12.1						1	1
12.3	1	1	1	1			
13.6	1	1	1	1	1		
14.3	1	1	1		1	1	
15.6	1				1	1	1
16.1	1	1			1	1	1
17.1	1	1	1	1	1		
18.1							1
18.3	1	1	1	1	1		
19.1	1	1			1	1	
20.3						1	
20.6	1		1	1			
21.1							1
21.3	1	1	1	1	1		
22.3	1			1	1		
22.6						1	
23.1	1			1			
23.3						1	1
24.3	1	1	1	1	1		
25.1	1	1	1	1			
25.6						1	1
26.1		1	1	1	1		
26.3							1
27.1						1	1
27.6		1	1				
29.1		1	1	1	1	1	
Type 2 (LPN)							
1.6						1	1
2.2		1	1	1	1		
2.6						1	
3.4					1		

(continued)

Appendix 4 (continued)

Week 1							
(ID.Shift)	Monday	Tuesday	Wednesday	Thursday	Friday	Saturday	Sunday
3.5						1	1
4.5	1	1	1	1	1		
5.3		1					
5.4						1	1
6.3							1
6.6	1				1		
7.7				1	1	1	1
8.2	1	1		1	1		
9.2	1		1		1		1
10.5	1	1			1		
10.6						1	
11.2						1	1
11.4				1			
Type 3 (NA)							
1.3						1	
1.5			1	1			
2.5	1	1	1	1	1	1	1
3.5						1	1
3.7	1	1	1				
4.2						1	
Type 4 (HCT)							
1.2				1		1	1
2.2	1	1	1				
2.6							1
3.4	1	1	1				
Week 3							
	Monday	Tuesday	Wednesday	Thursday	Friday	Saturday	Sunday
Type 1 (RN)							
1.3	1	1	1	1	1		
2.1							1
2.6					1		
3.3				1	1		
4.6	1		1	1	1	1	
5.1		1	1	1	1		
6.1	1	1		1	1	1	1
7.1		1		1	1	1	1
8.6	1	1		1	1		1
9.1	1		1	1	1		
9.6							1
10.3						1	1
10.6		1		1			
11.6	1			1	1	1	1
12.3						1	1
12.6		1	1	1			
13.3						1	1

(continued)

Appendix 4 (continued)

Week 1							
(ID.Shift)	Monday	Tuesday	Wednesday	Thursday	Friday	Saturday	Sunday
13.6	1	1	1				
14.1	1	1			1	1	1
15.1		1	1	1	1		
15.6						1	
16.1	1	1	1	1			
16.3						1	1
17.1	1	1	1	1			
17.6							1
18.1	1		1	1		1	1
19.3						1	1
19.6	1		1	1			
20.1	1	1	1		1	1	1
21.6		1	1	1	1	1	
22.1							1
22.3	1	1	1	1	1		
23.6	1	1	1		1		
24.3	1			1	1		
24.6						1	1
25.3		1	1		1	1	1
26.1						1	1
26.3	1	1	1				
27.3	1	1	1	1	1		
29.1						1	
29.3	1	1	1	1			
Type 2 (LPN)							
1.2				1			
2.7					1		
3.5						1	1
4.2	1	1	1	1			
4.5							1
5.3		1		1	1		
6.5		1	1	1	1	1	1
8.2						1	1
8.4	1		1	1			
9.5	1	1	1	1	1		
10.2	1	1			1	1	
11.2	1	1					
11.7						1	1
Type 3 (NA)							
1.2			1	1	1	1	
2.2							1
2.5	1		1				
3.2							1
3.4		1			1		
4.4						1	1
4.7	1	1	1	1			

(continued)

Appendix 4 (continued)

Week 1							
(ID.Shift)	Monday	Tuesday	Wednesday	Thursday	Friday	Saturday	Sunday
Type 4 (HCT)							
1.5	1	1		1	1		
2.2			1		1		
2.5						1	
3.3	1		1		1		
Week 4							
	Monday	Tuesday	Wednesday	Thursday	Friday	Saturday	Sunday
Type 1 (RN)							
1.1	1			1			
1.3						1	1
2.6		1			1	1	1
3.6							1
4.1							1
4.3		1	1	1	1		
5.1	1	1		1	1		
5.3						1	1
6.3							1
6.6	1		1		1		
7.1	1	1			1		
7.3						1	1
8.1						1	1
8.6	1	1		1			
9.6		1		1	1	1	1
10.3	1	1	1	1	1		1
11.1			1	1	1	1	1
12.3	1	1	1	1	1		
13.3	1	1	1		1	1	
14.1						1	
14.6	1	1	1	1			
15.1	1	1	1		1		
15.6						1	
16.1						1	1
16.6			1	1			
17.1		1	1	1	1		
17.6							1
18.1	1	1	1		1		
18.6						1	
19.1						1	1
19.6	1	1	1				
20.3							1
20.6	1		1		1		
21.6	1	1	1	1	1		
22.1	1	1	1	1			
23.3	1	1	1	1	1	1	

Appendix 4 (continued)

Week 1							
(ID.Shift)	Monday	Tuesday	Wednesday	Thursday	Friday	Saturday	Sunday
24.1		1	1	1		1	1
25.1						1	1
25.3	1	1		1			
26.1	1	1	1	1	1		
27.3	1	1	1	1	1		
28.6				1	1	1	1
29.3	1		1	1	1	1	
Type 2 (LPN)							
1.2	1			1		1	
2.5	1	1	1	1		1	1
3.5	1	1	1	1	1		
4.7						1	1
5.5	1	1	1	1	1		
6.4				1			
8.2						1	
8.4		1					
9.5					1		1
10.2						1	1
10.7					1		
11.2	1				1		1
Type 3 (NA)							
1.2							1
1.7		1	1				
2.2	1	1	1				
2.5						1	
3.2		1	1	1	1		
4.2					1		
Type 4 (HCT)							
1.5							1
1.7	1			1			
2.2		1	1	1			
2.5						1	
3.4	1		1		1	1	1
							1

References

1. Abernathy, W., Baloff, N., Hershey, J., Wandel, S.: A three-stage manpower planning and scheduling model-a service-sector example. Oper. Res. **21**(3), 693–711 (1973)
2. Beliën, J., Demeulemeester, E.: A branch-and-price approach for integrating nurse and surgery scheduling. Eur. J. Oper. Res. **189**(3), 652–668 (2008)
3. Burke, E.K., Bb N.N., De Causmaecker, P.: Fitness evaluation for nurse scheduling problems. Proceedings of the 2001 IEEE Congress on Evolutionary Computation, Seoul, Korea, pp. 1139–1146 (2001)
4. Burke, E., De Causmaecker, P., Berghe, G., Van Landeghem, H.: The state of the art of nurse rostering. J. Sched. **7**, 441–499 (2004)
5. Cheang, H., Lim, A., Rodrigues, B.: Nurse rostering problems-a bibliographic survey. Eur. J. Oper. Res. **151**, 447–460 (2003)
6. Chern, C.-C., Chien, P.-S., Chen, S.-Y.: A heuristic algorithm for the hospital health examination scheduling problem. Eur. J. Oper. Res. **186**(3), 1137–1157 (2008)
7. ILOG CPLEX 11.0 Users Manual, 2007.
8. Duenas, A., Tutuncu, G.Y., Chilcott, J.B.: A genetic algorithm approach to the nurse scheduling problem with fuzzy preferences. IMA J. Manag. Math. **20**, 369–383 (2009)
9. Ferland, J., Berrada, I., Nabli, I., Ahoid, B., Michelon, P., Gascon, V., Gagne, E.: Generalized assignment type goal programming problem: application to nurse scheduling. J. Heuristics **7**, 391–413 (2001)
10. De Grano, M.L., Medeiros, D.J., Eitel, D.: Accommodating individual preferences in nurse scheduling via auctions and optimization. Health Care Manage Sci. **12**, 228–242 (2009)
11. Gutjahr, W.J., Rauner, M.S.: An ACO algorithm for a dynamic regional nurse-scheduling problem in Austria. Comput. Oper. Res. **34**(3), 642–666 (2007)
12. Jaumard, B., Semet, F., Vover, T.: A generalized linear programming model for nurse scheduling. Eur. J. Oper. Res. **107**, 1–18 (1998)
13. Mohan, S.: Scheduling part-time personnel with availability restrictions and preferences to maximize employee satisfaction. Math. Comput. Model. **48**(11–12), 1806–1813 (2008)
14. Pardalos, P.M., Resende, M.: Handbook of Applied Optimization, Oxford University Press (2002).
15. Topaloglu, S., Selim, H.: Nurse scheduling using fuzzy modeling approach Fuzzy Sets and Systems, Corrected Proof, Available online 15 Oct 2009 (in press).
16. Tsai, C.-C., Li, S.H.A.: A two-stage modeling with genetic algorithms for the nurse scheduling problem. Expert Syst. Appl. **36**(5), 9506–9512 (2009)
17. Vanhoucke, M., Maenhout, B.: On the characterization and generation of nurse scheduling problem instances. Eur. J. Oper. Res. **196**(2), 457–467 (2009)
18. Warner, D.M.: Scheduling nursing personnel according to nursing preference: A mathematical programming approach. Oper. Res. 24:842–856, 1976
19. Wright, P.D., Bretthauer, K.M., Cote, M.J.: Reexamining the nurse scheduling problem: staffing ratios and nursing shortages. Decis. Sci. **37**(1), 29–70 (2006)

Clinical Data Mining to Discover Optimal Treatment Patterns

Patricia Cerrito[†]

1 Introduction

Medical data have some unique characteristics that must be considered in any type of analytics investigation. Generally, the customer is the decision maker, the payer, and the recipient of goods and services. In healthcare, the physician makes the treatment decisions, the insurer pays the bill, and the patient receives the goods and services. Because it is sometimes difficult to identify the customer in the healthcare system, inputs into models have to consider the different players. In addition, the final product is difficult to define. Is the outcome improved quality of life, the elimination of disease, or reduced cost?

There are many advantages and a few disadvantages in using large, observational databases to investigate healthcare outcomes. Currently, randomized, controlled trials remain the "gold standard" in health outcomes. Moreover, such studies typically enroll only patients at high risk. However, such studies are expensive and tend to rely upon surrogate endpoints and short-term results, so longitudinal, long-term outcomes are rarely examined. Observational studies can be used to examine actual outcomes and long-term results as well as investigate outcomes for those at more moderate risk. Such studies can also be used to investigate in detail the patient's perspective, which is extremely useful when examining patient quality of life, which is used to define quality adjusted life years in comparative effectiveness analysis.

Because the collected data are usually in multiple tables and datasets, a considerable amount of preprocessing is required. In particular, one-to-many relationships need to be changed to one-to-one for predictive modeling. Therefore, we first examine some of the preprocessing required for health outcomes analysis using data mining tools.

P. Cerrito[†]
(Deceased)

2 Preprocessing Data

In claims data, prescriptions are separated from inpatient and outpatient treatments as well as office visits and home health care. Because all of this information is stored in different files in a one-to-many relationship with a patient's identification number, the most important preprocessing need to use these databases is to convert them to a one-to-one relationship after filtering down to the condition under study. We take advantage of the data step and the use of summary statistics to do both. Each patient claim is identified by an ICD-9 code as to the primary reason for the medication or treatment. Osteoporosis, for example, is identified by the codes, 733.0x where x can vary from 0 to 9 (http://icd9cm.chrisendres.com/). Similarly, 153.xx identifies colon cancer. Each of the datasets has a column for the primary code, with space available for secondary diagnosis codes. We can use an if…then statement in a data step to isolate patients with a specific primary or secondary condition in order to investigate the major problem.

Once the different data sets have been filtered down to a specific condition, we need to convert them to a one-to-one relationship. There are many such databases in healthcare that have this one-to-many relationship. For example, a patient with a chronic condition will have many prescriptions per patient identifier. Similarly, there can be many physician visits, inpatient stays, or outpatient treatments for one patient. We then choose one of the datasets to serve as the primary set and merge the datasets using a left or a right join, depending upon the order of the data sets. In addition, we have to be concerned about whether a treatment type is discontinued, or if the patient switched to a different treatment type when performing the joins. We can also use summary information to create the join. While costs and charges can be summarized into a one-to-one relationship, other outcomes are not so easy, especially if the patient has a chronic disease that is progressive; we will need to examine the nature of the progression.

Because SAS software (SAS Institute, Inc.; Cary, NC) is used so commonly in medical research and drug development, we provide the SAS code for the preprocessing in our example in Sect. 6.

3 Analysis of Diagnosis and Procedure Codes for Clinical Data Mining

Probably the biggest issue in investigating healthcare databases is to identify the level of patient severity and to work with diagnosis codes that identify comorbidities, complications of treatment, and all potential patient procedures. Usually, just a small set of patient diagnoses are used to define the patient severity. This small set can create problems since diagnoses that have a higher probability of mortality or morbidity can be omitted from the set so that patients with some severe diagnoses are identified as not severe at all. These codes are of the form xxx.xx where the whole number is the general diagnosis and the decimal gets to specifics related to the major condition. There are thousands of these possible codes. In 2012, the healthcare system will start using ICD-10 codes, where there will be thousands more potential diagnoses.

Consensus panels are generally convened to decide upon a set of patient diagnoses to be used to define severity level. Generally, the panels have focused upon chronic diseases that are largely focused upon heart, cancer, breathing diseases such as asthma, and diabetes. Acute conditions are rarely used to define a severity level, including two conditions with very high mortality rates: immune disorder and septicemia.

These diagnosis codes represent nominal data or text information. Each patient can have an associated text string containing all listed diagnoses. Then, the text strings can be clustered using text analysis, which is very effective at clustering sequences of nouns. The clusters define a natural ranking of patient severity that can then be used in subsequent models. The procedure to do this is discussed at length in [1]. The process of text analysis of similarly identified text strings is demonstrated in Sect. 6 in our examples.

4 Predictive Modeling Versus Statistical Models

Observational databases of medical data tend to be quite large. Statistical models generally were designed to work with minimal and randomized data sets. The observational data used in health outcomes research are not randomized. Therefore, there is a serious question as to whether statistical models are sufficiently robust for health outcomes analyses. Because of the lack of randomness, potential confounding factors should be included in the model since large data sets allow for a larger number of variables in the model compared to randomized trials. One confounding factor will be the level of severity of the patient who likely will have comorbidities that may or may not be related to the primary condition studied in the model.

Suppose, for example, that a meta-analysis is performed using outcomes from several different studies. Suppose the purpose is to examine a rare occurrence such as mortality. Then, a generalized linear model with the outcome of mortality or other rare occurrence is required as opposed to a logistic regression model that requires an assumption of normality in the population. If the data are taken from several different studies, then a variable of "study" needs to be a random effect in the model since there are many studies available and only a handful are chosen to include in the model.

Moreover, the subjects must come from a random sample, which is defined as a sequence of independent and identically distributed random variables. If multiple studies are used in the model, each study clearly has different inclusion/exclusion requirements, especially including the type of disease, so the assumption of identical distributions is false; the subjects are not from a random sample. Because different studies targeted different diseases as identified by ICD9 codes, the specific disease type should be a nested variable inside of study. If nested effects are not used, the model is not valid.

The biggest problem with the model described above is that of predicting a rare occurrence of mortality or similar outcome regardless of whether the one study is used, or multiple studies. As discussed at length in [1], there has to be a link function for a Poisson distribution since a logistic regression model only works well if there is a 50/50 split in the outcome variable. Consider the linear regression model:

$$Y = \beta_0 + \beta_1 X_1 + \beta_2 X_2 \ldots + \beta_k X_k$$

where it is assumed that the residuals follow a normal distribution with mean zero and constant variance. When Y is dichotomous, a normal distribution is not possible. A standard logistic model then assumes the following model:

$$\log_e(p/1-p) = \beta_0 + \beta_1 X_1 + \beta_2 X_2 \ldots \beta_n X_n$$

where it can be assumed that the residuals are now normally distributed. However, if the value of p is small (or a count variable), then Poisson regression should be used as the Poisson distribution is a better approximation of the dichotomous variable than the normal approximation; the normal approximation requires p to be close to 50%:

$$\log_e(Y) = \beta_0 + \beta_1 X_1 + \beta_2 X_2 \ldots \beta_n X_n.$$

A better approach would be to use predictive models. There are usually components built into a predictive modeling approach that are extremely difficult using statistical models, especially with rare occurrences. With predictive modeling, prior probabilities can be assigned to give the true population proportion of the rare occurrence, and then the data set can be subsampled to a 50/50 split in the data and a logistic regression can be used. Moreover, weights can be assigned to differentiate between a false positive and a false negative since a false negative has a higher cost to a patient [2]. Then, instead of reporting odds ratios or relative risk, the actual misclassification rate can be used to find those patients who are at risk for the rare occurrence. In addition, multiple models can be used and compared to find the one with the highest level of accurate prediction.

It is possible to focus attention on those patients most at risk for a rare occurrence by using the concept of lift, which is very commonly used in data mining and predictive modeling; it is rarely used with statistical modeling. Using lift, true positive patients with highest confidence come first, followed by true positive patients with lower confidence. True negative cases with lowest confidence come next, followed by negative cases with highest confidence. Based on that ordering, the observations are partitioned into deciles, and the following statistics are calculated.

- The *Target density* of a decile is the number of actually positive instances in that decile divided by the total number of instances in the decile.

- The *Cumulative target density* is the target density computed over the first *n* deciles.
- The *lift* for a given decile is the ratio of the target density for the decile to the target density over all the test data.
- The *Cumulative lift* for a given decile is the ratio of the cumulative target density to the target density over all the test data.

Lift, then can find the 20% of patients most at risk rather than to consider everyone with a set of risk factors as being of risk. It enables practitioners to reduce costs by focusing on the highest risk patients. Standard logistic regression cannot distinguish between patients at high risk and those at moderate risk. For more information, we again refer the interested reader to Cerrito [1] and Cerrito [2].

5 Other Types of Data Mining

5.1 Market Basket Analysis

More specifically, association rules examine the strength of the treatment combinations that are used for a particular disease. An association rule is of the form $X \rightarrow Y$, meaning that X and Y are related such that if a patient has treatment X, then that same patient will generally have treatment Y. We can use association rules in a different way to examine relationships between patient conditions or between different treatments.

In addition to the antecedent X and the consequent Y, an association rule has two numbers that express the degree of uncertainty about the rule. In association analysis, the antecedent and consequent are sets of items that are disjoint ($X \cap Y = \emptyset$). The first number is called the support for the rule. It is the number of times that the combination appears. The support is simply the number of transactions in the denominator with all items in the antecedent and consequent parts of the rule in the numerator. The other number is known as the confidence of the rule. The confidence is the ratio of the number of transactions that include all items in the consequent as well as the antecedent to the number of transactions that include all items in the antecedent.

The support is equal to the number in common divided by the total number of transactions. The rules $X \rightarrow Y$ and $Y \rightarrow X$ can have different confidence values, but will have the same support values. The expected confidence is equal to the number of consequent transactions divided by the total number of transactions. The last measure of the strength of an association is the lift, which is equal to the ratio of the confidence to the expected confidence; that is, lift = confidence/expected confidence.

There are a number of ways to depict the results of a Market Basket analysis. The most visual is a link graph. There are nodes in the graph that depict possible targets. Links, called edges, connect nodes that are related. The size of the node depicts the number or support of the antecedent; the links connect an antecedent to a consequence. The thicker the link, the more the two are related. We will show several link graphs in Sect. 6.

5.2 Text Mining

Generally, a document is converted into a row in a matrix. This row has a column for any word contained within the data set of documents. The matrix value is equal to the number of times that word occurs in the document. The matrix will consist mostly of zeros since the list of words is much longer than the list of documents. Therefore, the next step is to reduce the dimension of the matrix. This is done through the process of singular value decomposition (SVD). Text analysis is extremely valuable for calls into customer service, for example, for chart notes, and to examine advertisements from the competition. It is extremely useful when examining patient comorbidities [1] and patient quality of life.

There are variations to this general methodology depending upon what you want to discover. For example, if you want to determine what documents contain a specific word for flagging purposes, this can be done through filtering. However, if you want to look at connections within the text structure itself, you can find much greater meaning using the word structure and natural language processing. The basics of text analysis are as follows.

1. Transpose the data so that the observational unit is the identifier and all nominal values are defined in the observational unit.
2. Tokenize the nominal data so that each nominal value is defined as one token.
3. Concatenate the nominal tokens into a text string such that there is one text string per identifier. Each text string is a collection of tokens.
4. Use text mining to cluster the text strings so that each identifier belongs to one cluster.
5. Use the clusters defined by text mining in other statistical analyses.

The general process of text analysis is outlined below.

The SVD of an $N \times p$ matrix A having N documents and p terms is equal to $A = U\Sigma V$, where U and V are $N \times p$ and $p \times p$ orthogonal matrices, respectively. U is the matrix of term vectors and V is the matrix of document vectors; Σ is a $p \times p$ diagonal matrix with diagonal entries $d_1 \geq d_2 \geq \ldots \geq d_p \geq 0$, called the singular values of Σ. The truncated decomposition of A is when the SVD calculates only the first K columns of U, Σ, and V. The SVD is the best least squares fit to A. Each column (or document) in A can be projected onto the first K columns of U. Similarly, each row (or term) in A can be projected onto the first K columns of V. The columns projection (document projection) of A is a method to represent each document by K distinct concepts. So, for any collection of documents, SVD forms a K dimensional subspace that is a best fit to describe the data.

Cluster analysis, also called data segmentation, has a variety of goals. All goals relate to grouping, or segmenting a collection of objects into subsets or "clusters" such that those elements within each cluster are more closely related to one another than the objects assigned to different clusters. An object can be described by a set of measurements or by its relation to other objects.

In addition, another goal is to arrange the clusters into a natural hierarchy. The arranging involves successively grouping the clusters themselves so that at each level of the hierarchy, clusters within the same group are more similar to each other than those in different groups. Cluster analysis is used to form descriptive statistics to assess whether or not the data consist of a set of distinct subgroups; each subgroup represents objects with substantially different properties.

Central to all of the goals of cluster analysis is the notion of the degree of similarity or dissimilarity between the individual objects being clustered. A clustering method attempts to group the objects based on the definition of similarity supplied to it. Clustering algorithms fall into three distinct types: combinatorial algorithms, mixture modeling, and mode seeking. In text analysis, we use mixture modeling.

Text analysis has as its basis the Expectation Maximization Algorithm. The expectation maximization (EM) algorithm uses a different approach to clustering in two important ways.

1. Instead of assigning cases or observations to clusters to maximize the differences in means for continuous variables, the EM clustering algorithm computes probabilities of cluster memberships based on one or more probability distributions. The goal of the clustering algorithm is to maximize the overall probability or likelihood of the data, given the final clusters.
2. Unlike the classic implementation of k-means clustering, the general *EM* algorithm can be applied to both continuous and categorical variables.

The EM algorithm is used to estimate the probability density of a given set of data. EM is a statistical model that makes use of the finite Gaussian mixtures model and is a popular tool for simplifying difficult maximum likelihood problems. The algorithm is similar to the K-means procedure in that a set of parameters is recomputed until a desired convergence value is achieved. The finite mixture model assumes all attributes to be independent random variables.

5.3 Comparative Effectiveness Analysis

The National Health Service in Britain has been using comparative effectiveness analysis for quite some time. NICE stands for the National Institute for Health and Clinical Excellence. This organization has defined an upper limit on treatment costs, and if the cost exceeds this pre-set limit, then the treatment is denied. It does not matter if the drug is effective or not. That means that there are many beneficial drugs that are simply not available to patients in Britain where fully 25% of cancer patients are denied effective chemotherapy medications [3, 4]. The number of chemotherapy drugs denied is increasing regardless of their effectiveness.

These drugs are denied by inflating the cost through adjustments for quality of life and by setting a threshold value to deny the treatment if the quality adjustment exceeds the threshold value. It is the researchers who define the patient's quality of life without actually considering the patient's own definition. Yet a recent survey showed that patients completely paralyzed remain content with their lives [5].

Instead of looking at patient data concerning the actual complaints from the drug treatment from clinical trials, comparative effectiveness analysis uses three general methods.

- Time Trade Off (TTO): Respondents are asked to choose between remaining in a state of ill health for a period of time, or being restored to perfect health but having a shorter life expectancy.

This trade off can be considered a "happy pill" guaranteed to give you perfect health for a period of time after which you drop dead while you can have twice as long to live but you have to put up with a chronic illness. As this "happy pill" does not exist, respondents are given a hypothetical, nonexistent choice. Moreover, this choice is usually asked of healthy people.

- Standard gamble (SG): Respondents are asked to choose between remaining in a state of ill health for a period of time, or choosing a medical intervention, which has a chance of either restoring them to perfect health or killing them.

While the respondents often are asked this as a hypothetical choice, patients often make this choice on a regular basis, including cancer and heart disease. Very few patients, when faced with this actual choice, actually opt to avoid the treatment. However, comparative effectiveness analysis often disallows a choice by rationing the treatment for older age groups, or those in need of expensive medical interventions.

- Visual Analogue Scale(VAS): Respondents are asked to rate a state of ill health on a scale from 0 to 100, with 0 representing death and 100 representing perfect health.

This is a question that can be asked of healthy individuals as well as those who have a particular state of ill health.

In the next section, we look at voluntary complaints concerning the use of the drug, Avastin, as well as other drugs used to treat colon cancer. We can use this information to generate hypotheses that can be examined using treatment databases. We demonstrate how actual data can be used to investigate the patient perspective, rather than to use hypothetical situations and responses given by healthy people.

6 Example Analysis of Colon Cancer Adverse Events

We look to the AERS database located at the web site, http://www.fda.gov/Drugs/GuidanceComplianceRegulatoryInformation/Surveillance/AdverseDrugEffects/ucm082193.htm. This web site is sponsored by the Centers for Disease Control and contains information concerning adverse events. Because the reporting is voluntary, it is not known if these reports are valid. However, it represents a starting point to examine healthcare information using data mining techniques. There are multiple datasets defined by one quarter of the year. Four quarters are readily available for download; datasets for the years starting from 2004 can also be downloaded. In the information accompanying the datasets, the following files are available.

1. DEMOyyQq.TXT contains patient demographic and administrative information, a single record for each event report
2. DRUGyyQq.TXT contains drug/biologic information for as many medications as were reported for the event (1 or more per event)
3. REACyyQq.TXT contains all "Medical Dictionary for Regulatory Activities" (MedDRA) terms coded for the event (1 or more). For more Information on MedDRA, please contact: TRW, VAR 1/6A/MSSO, 12011 Sunset Hills Road, Reston, VA 20190-3285, USA; web site is http://www.meddramsso.com
4. OUTCyyQq.TXT contains patient outcomes for the event (0 or more)
5. RPSRyyQq.TXT contains report sources for event (0 or more)
6. THERyyQq.TXT contains drug therapy start dates and end dates for the reported drugs (0 or more per drug per event)
7. INDIyyQq.TXT contains all "Medical Dictionary for Regulatory Activities" (MedDRA) terms coded for the indications for use (diagnoses) for the reported drugs (0 or more per drug per event)

We also use the MedDRA coding to translate into English the adverse events listed. We can then define text strings to investigate the complaints. We can also use market basket analysis.

6.1 Treatment Investigated

First-line treatment for colorectal cancer generally consists of three drugs: 5-fluorouracil, folinic acid, and oxaliplatin. The three drugs in combination are abbreviated as FOLFOX. Avastin is added to this combination for metastatic colon cancer. If irinotecan is substituted for the oxaliplatin, the combination is identified as FOLFIRI. We will examine the first-line treatment of FOLFOX in this analysis. We will include medications often administered to help to alleviate the side effects of the chemotherapy treatment.

6.2 Comparative Effectiveness Analysis for Colon Cancer Treatments

While comparative effectiveness models use the results of previously published studies, they do not use any observational data. We want to improve upon the models through the use of data mining to examine optimal treatment patterns. Although we want to examine all of the drugs in the data, we will focus the analysis on the potential adverse events of a drug, Avastin (bevacizumab), that has come under attack in both Britain and the USA using comparative effectiveness analysis and benefit/risk ratio.

In the USA, the drug has been approved for several types of cancer; however, it has recently been delisted (or approval withdrawn) for breast cancer. While there is no good evidence that the drug improves overall survival for breast cancer, there is evidence that it improves disease-free survival. It was withdrawn because disease-free survival was deemed an insufficient benefit compared to the drug risks [6,7]. At the same time, there are many patients with breast cancer who are taking the drug because they themselves believe that the benefits do outweigh the risks.

In Britain, the first report produced by NICE that examined the drug, bevacizumab (Avastin) was analyzed based upon a complete lack of knowledge of a patient's quality of life. Because of the lack of availability of such data, the model assumed that there was no improvement in quality from the drug [8]. Given that the drug is quite effective for colon cancer and has moderate side effects compared to other drugs, that assumption is a considerable stretch.

The drug was rejected because the cost effectiveness based upon an assumed additional 5 months of life was £39,136 to £69,439, beyond the limited threshold supported by the National Health Service, even though studies show much more than 24 months of life improvement with the drug, to the point that it is now considered as maintenance therapy [9–12]. The report was based upon two clinical trials only. The actual cost was less than half the adjusted price, and this actual cost falls under the threshold value imposed by NICE. Therefore, the use of quality adjusted life years (QALY) essentially inflates the actual cost, given that the QALY cost is not one that is actually paid.

Because the results of the comparative effectiveness model depend so heavily upon the definition of a patient's quality of life, we examine different databases that can be used to examine the definition of quality from the perspective of the patient with colorectal cancer.

7 Data Preprocessing

We use query and filter, and append capabilities that are available in SAS Enterprise Guide in order to combine information from the different datasets in the AERS databases in order to analyze the information. There is a one-to-many relationship in the data, and we need to find a way to define a one-to-one relationship between adverse events described and one voluntary report. We use the one-to-many relationship for market basket analysis while using the one-to-one relationship for predictive modeling. We change the data to one-to-one by transposing the information and then by concatenating it using the following SAS code.

PROC SORT DATA=AERS.colonchemo OUT=WORK.SORTED2;
by isr;
PROC TRANSPOSE DATA=WORK.SORTED2 OUT=AERS.TRANSPOSE2;
VAR INDI_PT;
by isr;
RUN;

```
data aers.concatcolon;
set aers.transpose2;
Adverse=catx(' ',col1, col2, col3, col4, col5, col6, col7, col8, col9, col10, col11,
col12, col13, col14, col15);
run;
```

We then limit the data set to the drugs generally identified as front-line treatment for colon cancer, along with additional medications that are administered to tolerate the chemotherapy treatment.

```
PROC SQL;
CREATE TABLE
SASUSER.FILTER_FOR_DRUG09Q1_0000 AS
SELECT t1.ISR,
t1.DRUG_SEQ,
t1.ROLE_COD,
t1.DRUGNAME,
t1.VAL_VBM,
t1.ROUTE,
t1.DOSE_VBM,
t1.DECHAL,
t1.RECHAL,
t1.LOT_NUM,
t1.EXP_DT,
t1.NDA_NUM
FROM SASUSER.DRUG09Q1 AS t1
WHERE t1.DRUGNAME IN ('AVASTIN', 'ALOXI', 'BEVACIAUMAB',
'BEVACIZUMAB',
'BEVACIZUMAB (INJECTION FOR INFUSION)', 'BEVACIZUMAB
(GENENTECH) - STUDY AGENT',
'BEVACIZUMAB (GENENTECH, INC.)', 'BEVACIZUMAB (GENETECH)
- STUDY AGENT',
'BEVACIZUMAB (INJECTION FOR INFUSION)', 'BEVACIZUMAB
(RHUMAB VEGF)', 'BEVACIZUMAB 10MG/ KG GENENTECH',
'BEVACIZUMAB 10MG/KG', 'BEVACIZUMAB 10MG/KG GENENTECH',
'BEVACIZUMAB 15 MG/KG',
'BEVACIZUMAB, TEST ARTICLE IN TEMSIROLIMUS STUDY',
'DECADRON',
'DECADRON /CAN/', 'DECADRON /NET/',
'DECADRON #1', 'DECADRON (DEXAMETHASONE PHOSPHATE)',
'DECADRON /00016002/', 'DECADRON /NET/', 'DEXA',
'DEXAMED', 'DEXAMETASON', 'DEXAMETHASONE',
'DEXAMETHASONE SODIUM PHOSPHATE', 'DEXAMETHASONE
SODIUM POSPHATE',
'DEXAMETHASONE TAB', 'ELOXATIN', 'FLUOROURACIL', 'FLUOROU
RACIL+CYCLOPHOSPHAMIDE', 'FLUOROURACILE TEVA',
```

Table 1 Summary of first-line colon cancer treatment

Drug	Frequency	Percent
5FU	4,671	20.08
Aloxi	496	2.13
Avastin	6,091	26.19
Decadron	9,843	42.32
Leucovorin	1	0.00
Oxaliplatin	2,155	9.27

Table 2 Primary indications for drug use

Primary indication	Frequency	Percent
Drug use for unknown indication	826	21.46
Multiple myeloma	328	8.52
Colorectal cancer	164	4.26
Breast cancer	160	4.16
Colon cancer metastatic	122	3.17
Premedication	119	3.09
Colorectal cancer	85	2.21
Non-small cell lung	83	2.16
Weight loss diet	74	1.92
Breast cancer metastatic	66	1.71
Chemotherapy	63	1.64
Prophylaxis	56	1.45
Pain	48	1.25
Obesity	44	1.14
Glioblastoma multiforme	41	1.07
Weight control	39	1.01
Gastric cancer	37	0.96
Rectal cancer	37	0.96

'LEVCOVORIN', 'OXAIPLATIN', 'OXALIPLATIN', 'OXALIPLATINE', 'PALONOSCTRON', 'PALONOSETRON');
QUIT;

The next step is to examine and summarize the information in the database. There are a total of 24, 366 different complaints for 4,278 individuals. Table 1 summarizes the medications listed in the complaints.

Decadron is a steroid used for many different conditions so the fact that it accounts for almost half of the complaints can be misleading; it is prescribed more often compared to the other drugs. Avastin also has a large number of complaints since it is used for multiple types of cancer. Table 2 shows some of the conditions stated as reasons for taking the drugs.

Table 2 shows some of the problems in examining the reason for taking the medications. Note that colorectal cancer is number 5, but colon cancer is number 7, and metastatic colon cancer is number 9 on the list. It shows that there can be slight differences in the wording that can lead to multiple categorical levels that should be

Table 3 Primary indications limited to Avastin

Primary indication	Frequency	Percent
Drug use for unknown indication	2,383	39.49
Breast cancer	399	6.61
Non-small cell lung	350	5.80
Colon cancer	266	4.41
Colorectal cancer metastatic	216	3.58
Breast cancer metastatic	196	3.25
Colorectal cancer	167	2.77
Colon cancer metastatic	134	2.22
Rectal cancer	102	1.69
Metastatic renal cell carcinoma	83	1.38
Hypertension	81	1.34
Glioblastoma multiforme	76	1.26
Renal cell carcinoma	74	1.23
Neoplasm malignant	66	1.09
Premedication	66	1.09
Ovarian epithelial cancer	63	1.04
Pain	50	0.83
Large intestine carcinoma	47	0.78
Macular degeneration	47	0.78
Oesophageal carcinoma	38	0.63
Prophylaxis	32	0.53
Nausea	29	0.48
Lung neoplasm malignant	28	0.46
Rectal cancer recurrent	27	0.45
Metastases to liver	25	0.41
Ovarian cancer	23	0.38
Colon cancer recurrent	22	0.36
Pancreatic carcinoma metastatic	22	0.36
Rectal cancer metastatic	22	0.36

reduced to just one level. In addition, there are multiple columns that can give many different reasons for taking the drugs. We also want to know if there is a difference in complaints for the individual drugs. Since we also want to focus on the drug, Avastin, Table 3 shows the most frequent primary indications for that drug.

Because we want to concentrate on colon and breast cancer, we need to start combining these drug indications. We do this by concatenating the indications into one text string and then using the 'contains' feature in SAS preprocessing. We use the following code to isolate colon cancer and then use similar code for breast cancer. We then append the two files that are created. There are a total of 351 reports for breast cancer and 872 reports for colon cancer.

PROC SQL;
CREATE TABLE SASUSER.QUERY_FOR_COLONSTART_SAS7BDAT AS
SELECT t1.ISR,
t1.DRUG_SEQ,

```
t1.ROLE_COD,
t1.DRUGNAME,
t1.VAL_VBM,
t1.ROUTE,
t1.DOSE_VBM,
t1.DECHAL,
t1.RECHAL,
t1.LOT_NUM,
t1.EXP_DT,
t1.NDA_NUM,
t1.modified_drug_name,
t1.ISR1,
t1._NAME_,
t1.Adverse,
t1._DOCUMENT_,
t1._ROLL_1,
t1._ROLL_2,
t1._ROLL_3,
t1._ROLL_4,
t1.PROB1,
t1.PROB2,
t1.PROB3,
t1.PROB4,
t1.PROB5,
t1.PROB6,
t1._CLUSTER_,
/* CANCERTYPE */
('COLON') AS CANCERTYPE
FROM EC100024.colonstart AS t1
WHERE t1.Adverse CONTAINS 'COLON' OR t1.Adverse CONTAINS
'RECTAL' OR t1.Adverse CONTAINS 'LARGE INTESTINE' OR
t1.Adverse CONTAINS 'COLORECTAL';
QUIT;
```

As each quarter of data can have well over 300,000 indications, we first want to filter down to the reports for the colon and breast cancer patients. We will do this separately for each quarter and then append the data sets together. We can then examine the relationship of individual drug to individual complaint using market basket analysis and predictive modeling.

7.1 Market Basket Analysis

As the different complaints in the AERS database are listed by a report identifier, we can use market basket analysis to examine the connections between complaints. We can also examine the connections between drugs that are listed in the same

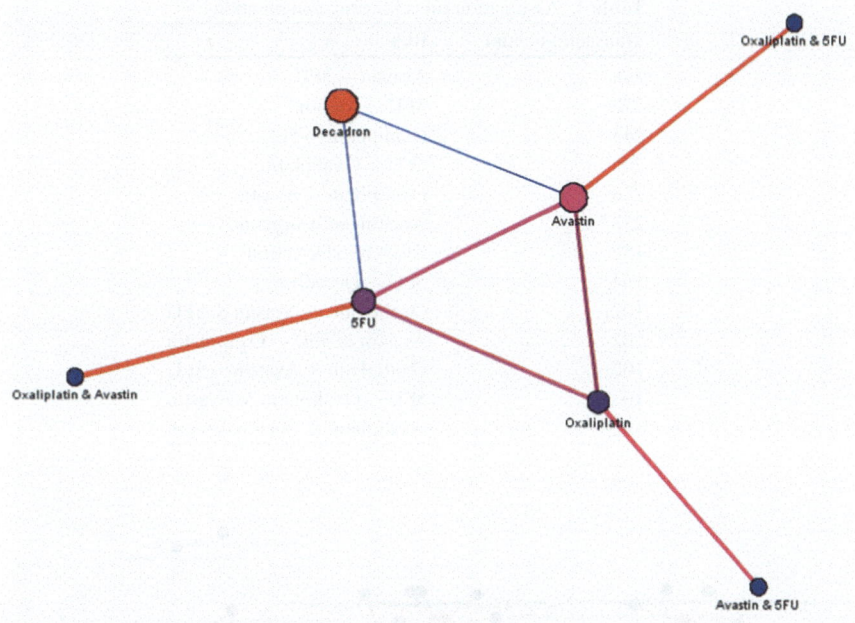

Fig. 1 Link graph of market basket analysis of voluntary complaints

report. To do so, we can use different target variables. Because we want to look at the total number of complaints that are represented in the database, we use the transaction count as the outcome to be examined in the link graphs.

Figure 1 shows the market basket analysis of complaints. It shows how complaints rather than treatments are linked together. It shows that Avastin is linked to both Oxaliplatin and 5FU, which suggests that it may be difficult to separate complaints concerning drugs that are given in combinations.

The link graph is a representation of the association rules concerning the combination of drugs in the complaints. To compare the graph to the rules, Table 4 gives these rules along with their transaction counts.

Both the link graph and the table clearly demonstrate that it is very difficult to separate out complaints for just the one drug, Avastin, since the drugs are given in combination and the adverse events are experienced by patients who are taking these combinations. Many of the complaints could be from chemotherapy generally rather than complaints from Avastin and just attributed to Avastin alone. We will investigate combinations versus single drugs in our market basket analysis.

We look now to the complaints using similar preprocessing commands as shown in Sect. 6.1. This includes merging to the data set that identifies the drugs listed using a left outer join. Figure 2 shows the overall complaints using a link graph. It shows four major centers and three minor centers. Since the overall graph in Fig. 2 is difficult to read other than to identify the number and size of patterns, Figs. 3, 4, 5, and 6 show the four major centers in greater detail.

Table 4 Association rules for drug combinations

Transaction count	Rule
278	Avastin → 5FU
278	5FU → Avastin
245	Oxaliplatin → 5FU
245	5FU → Oxaliplatin
238	Oxaliplatin → Avastin
238	Avastin → Oxaliplatin
152	Avastin → Decadron
104	5FU → Decadron
102	Oxaliplatin → Avastin & 5FU
102	Avastin & 5FU → Oxaliplatin
102	Oxaliplatin & Avastin → 5FU
102	5FU → Oxaliplatin & Avastin
102	Oxaliplatin & 5FU → Avastin

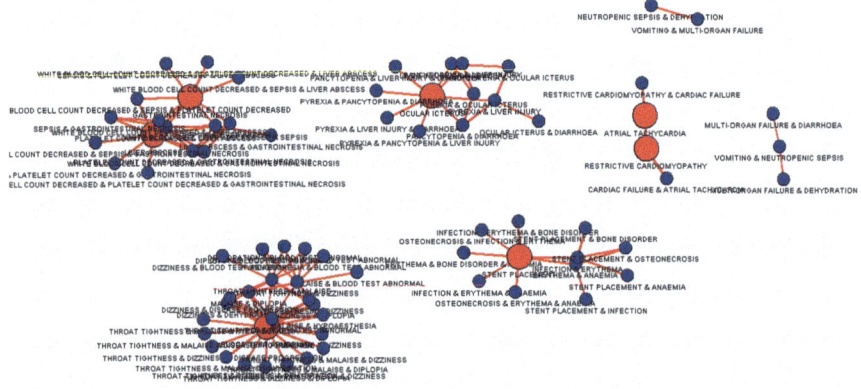

Fig. 2 Link graph of complaints for all medications

The first pattern deals mostly with low blood counts, particularly low platelet counts and throat tightness. These are not unusual side effects of chemotherapy generally [13].

The second pattern is for stent placement and infection; neither are unusual for medical procedures generally that require stent placement and can lead to infection because infection is prevalent in the environment of healthcare. These identified problems are not specific to chemotherapy, let alone to a specific drug. Infection might be more likely with chemotherapy because of the problem of low blood counts, and this connection suggests that Pattern 1 is related to Pattern 2 even though there is no apparent connection between the two patterns.

Figure 5 is very similar to the pattern shown in Fig. 3 as it has to do with cell counts and infection. However, the infection of sepsis is much more severe than just

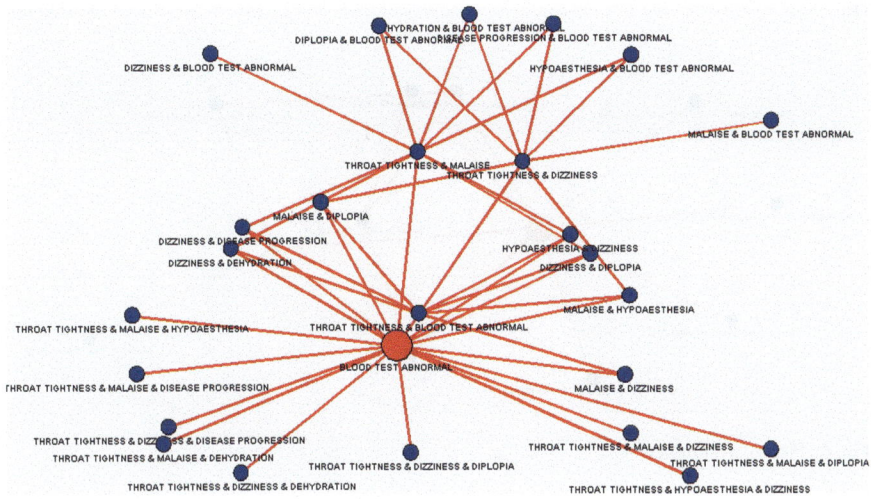

Fig. 3 Pattern 1-abnormal blood tests and throat tightness

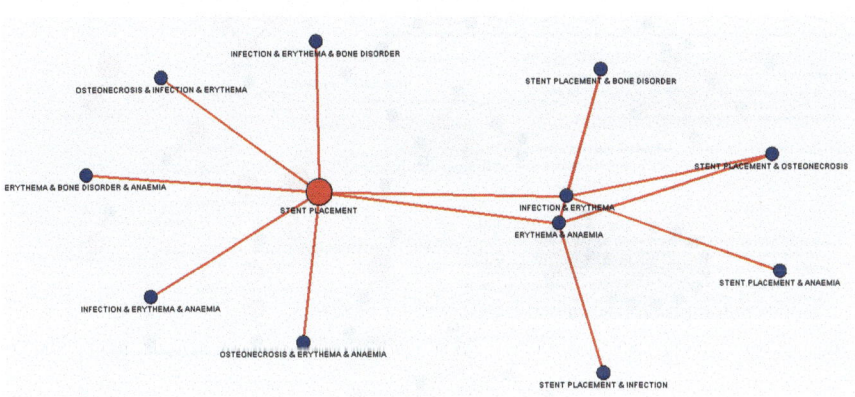

Fig. 4 Pattern 2-stent placement and infection

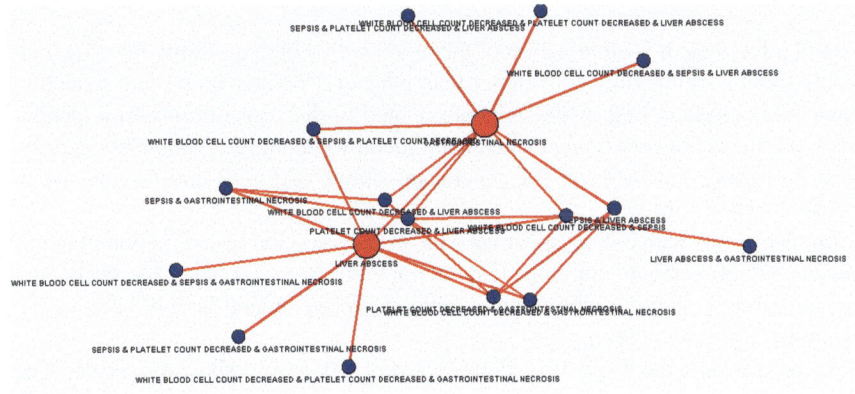

Fig. 5 Pattern 3-cell counts and sepsis

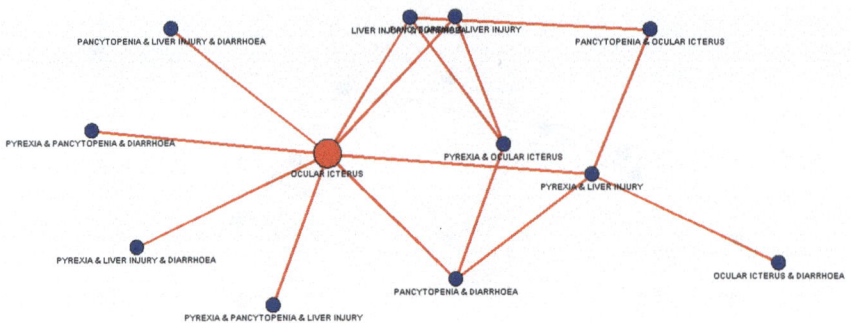

Fig. 6 Pattern 4-jaundice

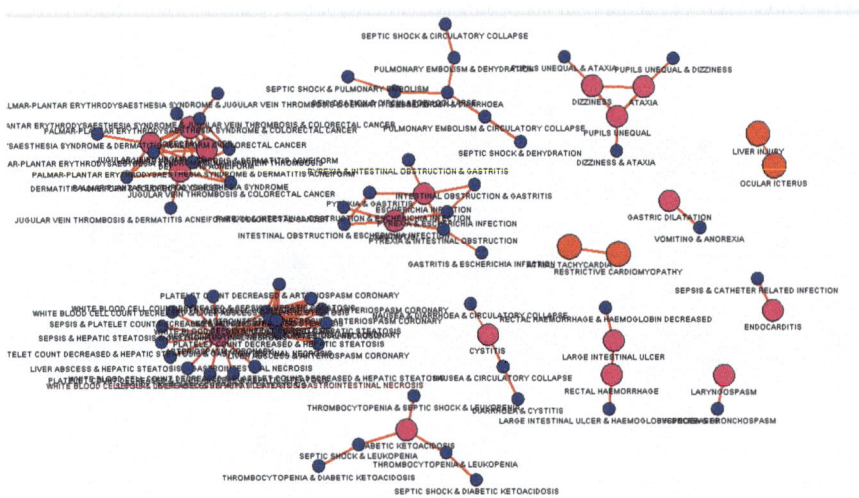

Fig. 7 Link graph of associated complaints for 5FU

general infections. It is an infection of the blood with a high mortality rate. The cell counts also focus on white cells rather than platelets. While platelets are related to injury and blood clotting, white cells are related to the susceptibility to infection. These problems are very typical with chemotherapy generally. Again, there appears to be no connection between the general infections of Fig. 3 and the low white counts shown in Fig. 5.

Many patients with colon cancer have metastasizes in the liver, so a complaint of jaundice is not so much from the chemotherapy but from a progression of the disease. It strongly suggests that these complaints are generated largely because the chemotherapy is not working.

We next isolate the drug, 5FU. Figures 7, 8, 9, 10, and 11 show the results. The graph in Fig. 7 shows two large centers and several smaller centers. Again, because

Clinical Data Mining to Discover Optimal Treatment Patterns 117

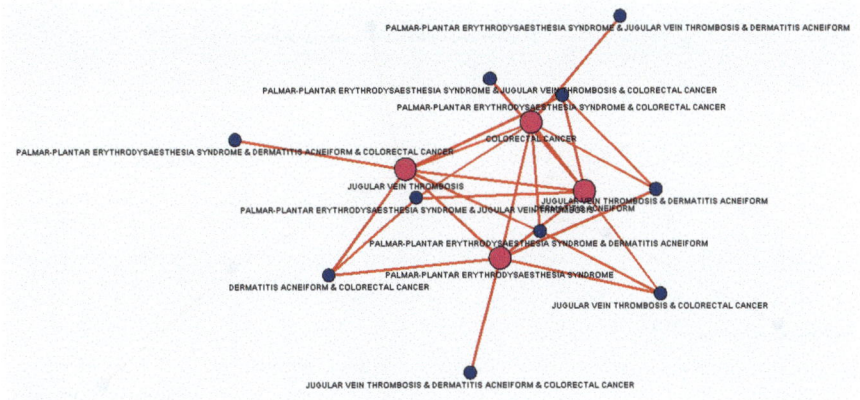

Fig. 8 Pattern 1 for 5FU-jugular vein thrombosis

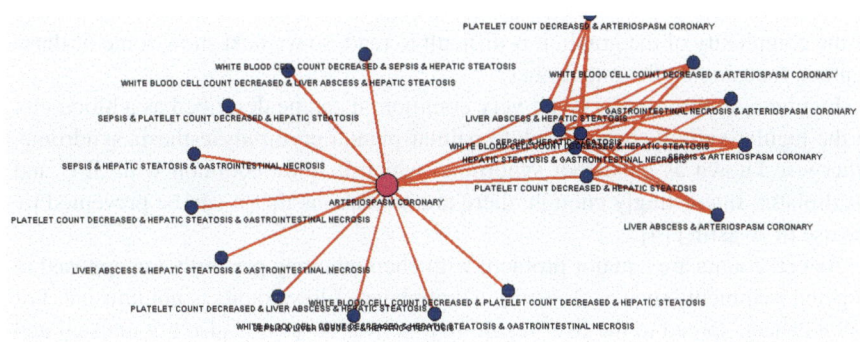

Fig. 9 Pattern 2 for 5FU-cell count and liver abscess

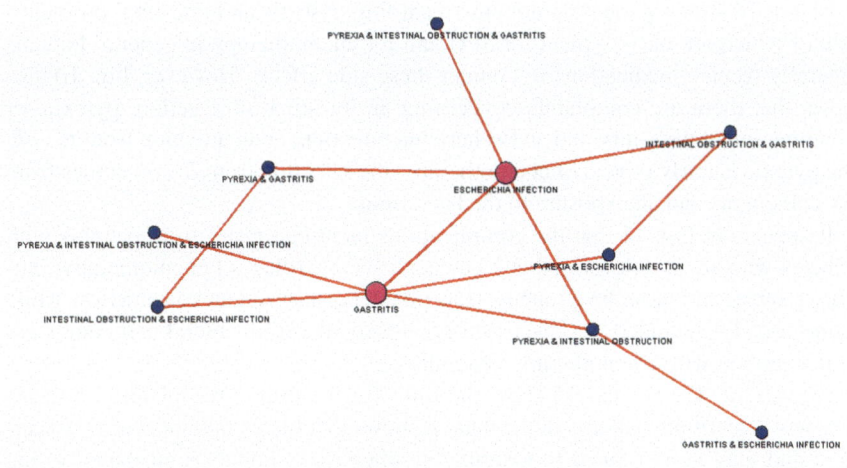

Fig. 10 Pattern 3 for 5FU-gastritis and intestinal problems

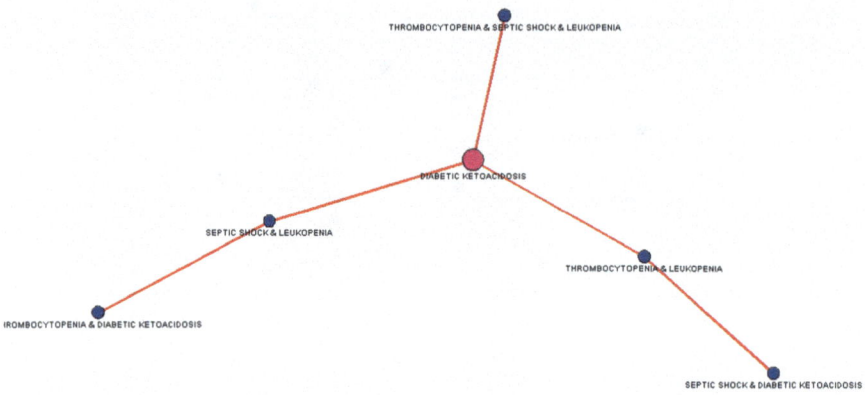

Fig. 11 Pattern 4 for 5FU-diabetes and severe complications

of the complexity of the graph, it is difficult to read, so we next show some of those centers, focusing on the major ones.

Jugular vein thrombosis is not very common; it can be described as a blood clot in the jugular vein. It shows a link to palmar-plantar erythrodysesthesia syndrome, otherwise known as hand-foot syndrome, which is quite common with 5FU and Oxaliplatin. Interestingly enough, there are indications that it can be prevented by the use of Avastin [14].

As cell counts are a major problem with chemotherapy generally (as opposed to targeted treatments such as Avastin), a complaint of low counts is not unusual. For colon cancer, spread to the liver resulting in a liver abscess is also not unusual and indicates a progression of the disease. These complications are also connected to an arteriospasm coronary and sepsis, with sepsis also identified as a complaint related to chemotherapy generally.

Figure 10 shows patient complaints regarding gastritis and intestinal problems, both of which are fairly typical for 5FU and for chemotherapy in general. Patients generally receive medications to counter these side effects. However, Fig. 10 also shows that there are complaints concerning an intestinal obstruction, pyrexia, or inflammation of the gums, and an Escherichia infection, or an infection from *E. coli*. The pyrexia is fairly typical of chemotherapy; the infection from *E. coli* comes from low cell counts and an exposure to the bacterium.

It appears in Fig. 11 that the complications identified result from patients with diabetes who are at risk for diabetic ketoacidosis regardless of chemotherapy; otherwise, sepsis and thrombocytopenia (low platelet count) and leukopenia (low white count) are also included. As discussed previously, these low blood cell counts are quite common with chemotherapy generally.

Figures 12, 13, 14, and 15 show the links for the drug, Oxaliplatin. There are very few discernible patterns in the data; it shows that the complaints are very scattered and may be attributed to Oxaliplatin when they should be attributed to the chemotherapy treatment in general. It also shows that there is no real pattern to the

Clinical Data Mining to Discover Optimal Treatment Patterns 119

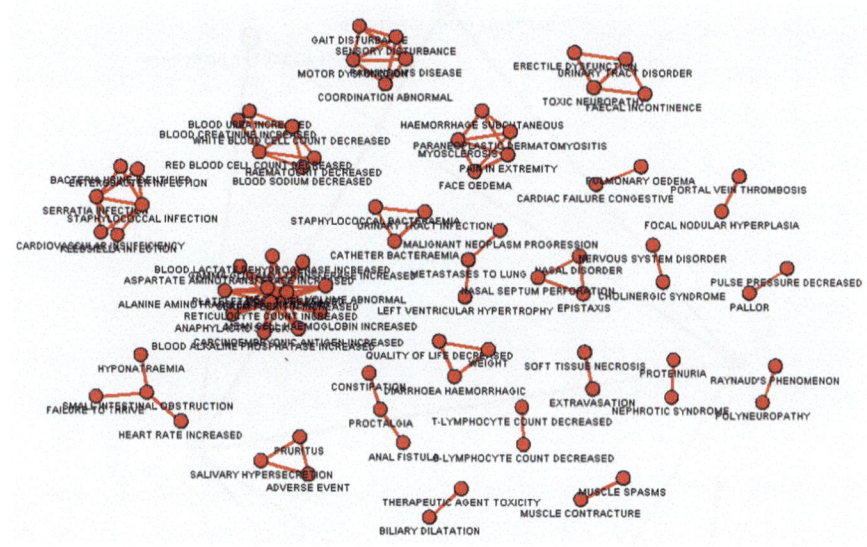

Fig. 12 Links connecting complications listed for Oxaliplatin

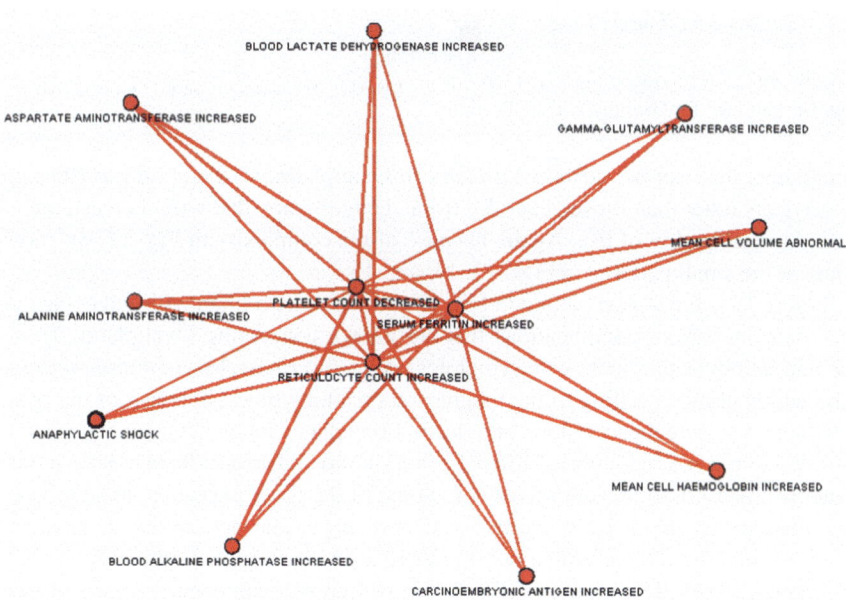

Fig. 13 Pattern 1 for Oxaliplatin complications

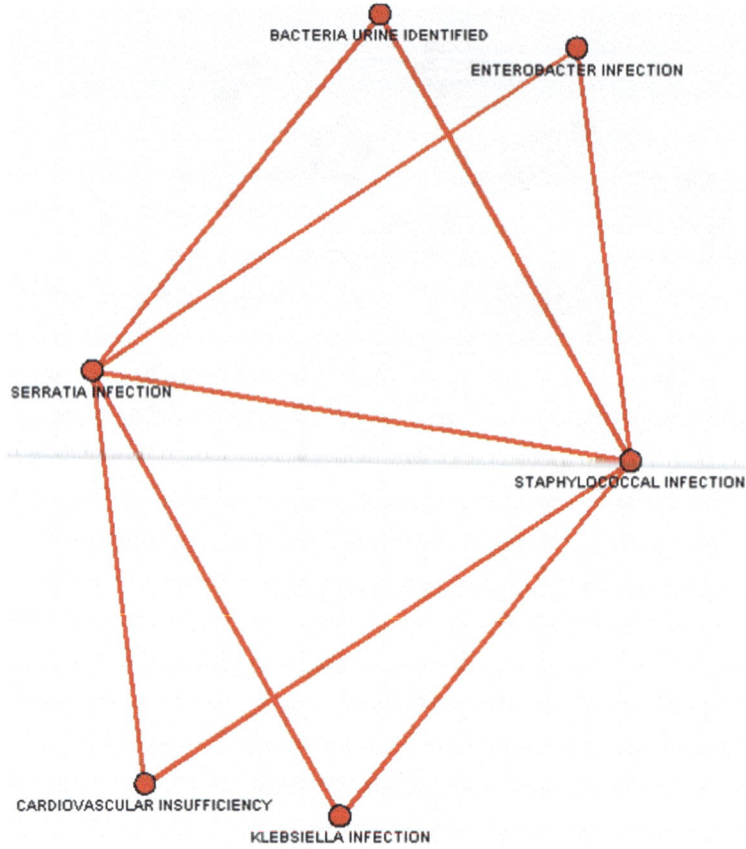

Fig. 14 Pattern 2 for Oxaliplatin complications

complaints that can be attributed directly to Oxaliplatin, so there are probably no side effects other than those generally from chemotherapy that would contribute to a diminished quality of life. Again, because of the complexity in Fig. 12, we show some of the small patterns in Figs. 13, 14, and 15.

Figure 13 is the largest pattern in Fig. 12 and shows a problem with cell counts as well as some severe complications that are known concerning Oxaliplatin. These include anaphylactic shock, with symptoms such as dizziness, loss of consciousness, labored breathing, swelling of the tongue and breathing tubes, blueness of the skin, low blood pressure, heart failure, and death. Immediate emergency care is required. Because of the risk, patients are often given a steroid and an anti-histamine as a preventative measure to prevent this shock. Some of the other measures relate to liver enzymes, which are elevated because of liver damage. For metastatic colon cancer, liver enzymes are often elevated because there are active tumors in the liver.

Figures 14 and 15 indicates the risk of infection that can occur because of low white cell counts, with several infections given specifically, including staph

Clinical Data Mining to Discover Optimal Treatment Patterns 121

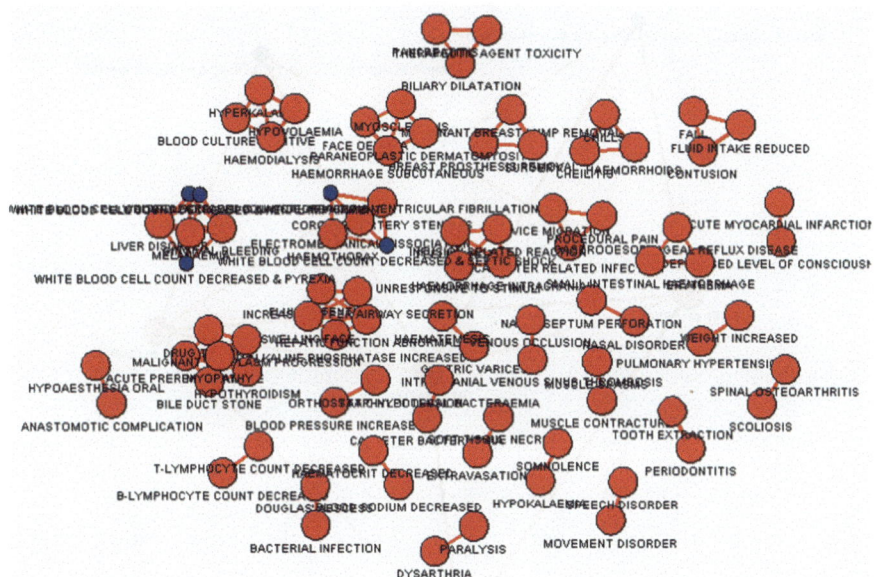

Fig. 15 Associations for Avastin

infection. It is not known if Oxaliplatin aggravates the problem of infection in terms of severity, or if the risk is generally related to chemotherapy.

Figures 14, 15, 16, 17, and 18 show the connections between complaints related to the drug, Avastin. Figures 14 and 15 clearly shows that there is no discernable pattern (similar to the results for Oxaliplatin) to the complaints, again suggesting that the complaints are related to chemotherapy generally rather than to just the one drug, Avastin. We will look at three of the small patterns to examine specifics.

Most of the complications are related to low cell counts, which are more likely from systemic chemotherapy rather than to targeted therapy such as Avastin (low cell counts, gingival bleeding, liver disorder). Two complications are related to Avastin and are known, renal impairment and melanaemia (The presence of dark brown or black granules of insoluble pigment in the blood), which are known problems with Avastin. Because of the possibility of renal impairment, blood and urine tests are usually performed prior to treatment to ensure that this complication does not happen.

The complications in Fig. 17 center on problems of the heart along with the white cell count problem that is known for chemotherapy in general. It suggests that these complications can be attributed to patients with heart disease of some type, including coronary artery stenosis.

Fluid retention and swelling face are likely (Fig. 18) because of the medications provided to deal with the side effects of the chemotherapy generally. Increased upper airway secretion is also a complication of chemotherapy generally. Abnormal hepatic function occurs in colon patients with metastasizes in the liver. Generally,

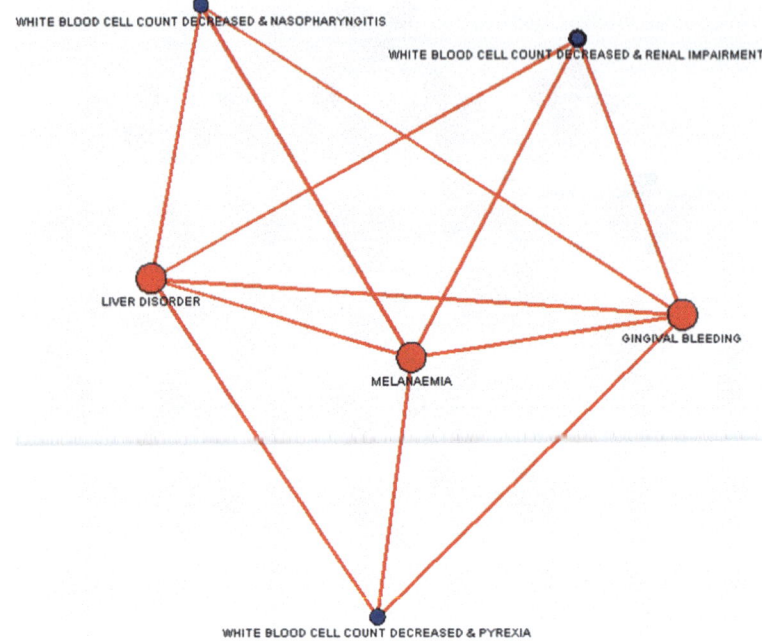

Fig. 16 Pattern 1 for Avastin complications

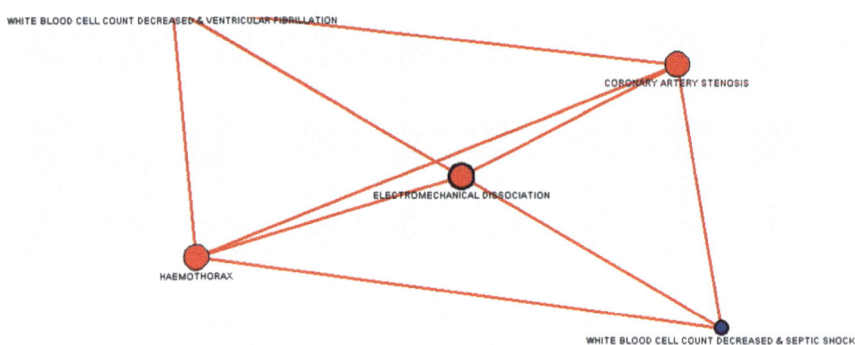

Fig. 17 Pattern 2 for Avastin complications

then very few of the problems with Avastin can be directly attributed to Avastin other than some mild bleeding. It suggests that adding Avastin to another regimen (FOLFOX, FOLFIRI) does not significantly enhance the side effects. Because it is rarely given alone, it should be considered as an add-on to that treatment when defining quality of life rather than as a stand-alone quality. However, the comparative effective analyses completed by NICE regard the quality of life separately for Avastin without looking at the marginal change due to adding Avastin to other potent drugs.

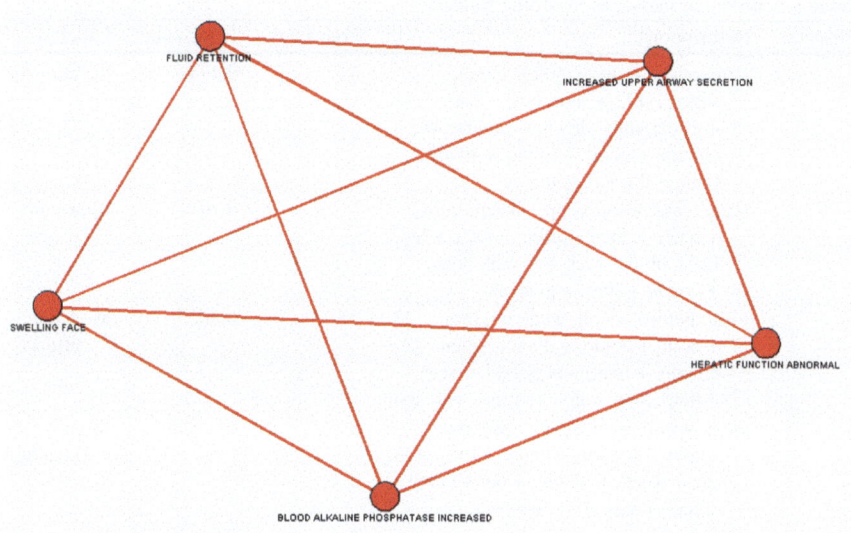

Fig. 18 Pattern 3 complications for Avastin

7.2 Text Mining

In this study, we used a total of 70 messages related to the use of the drug, Neulasta, which is frequently used for patients on chemotherapy. The drug has very potent side effects (listed at http://www.rxlist.com/neulasta-drug.htm). They include severe bone and muscle pain and the possibility of a ruptured spleen. The purpose of the medication is to grow white cells in cancer patients who have had white cells depleted because of chemotherapy treatments in order to prevent infections. It is often given to patients on FOLFOX and FOLFIRI along with Avastin, but the need cannot be attributed specifically to Avastin. As shown in Sect. 7.1, the chemotherapy has many complaints of low white cell counts.

The messages were written by patients or by caregivers of patients regarding their personal experiences. The messages are relatively short, but are related to quality of life. Interestingly, only two of the messages contained the word, "stop," indicating that most patients thought that the medication was worthwhile in spite of the problem of pain. We first examine the text clusters to see the most important terms involved in each group (Table 5).

Note that we have added tentative labels to each of the clusters. Labels are extremely useful when attempting to identify the most important relations in the clusters. Without labels, it is very difficult to find a use of these text clusters in subsequent analyses. Almost half of the messages are in cluster 1, which is characterized by the term, "severe." Another 36% indicate moderate pain in cluster 3. These messages give a strong sentiment that those who face difficult alternatives opt in favor of treatment as all but two of the messages indicate that treatment should be stopped. Table 6 is filtered to those messages that contain the word, "pain."

Table 5 Clusters of messages regarding Neulasta

Cluster #	Description	Frequency	Percentage	Label
1	+ injection, + agree, + side effect, severe, side, + hurt, but, + drug, + effect, only, down, + experience, chemo, + good, + round, + make, with, + do, + day, in	34	0.4927	Severe pain
2	bring, case, + week, up, + keep, same, + receive, back, during, much, + leg, little, blood, ever, + would, body, + count, + shot, + give, as	7	0.1014	Back pain
3	+ eat, before, + call, walk, no, into, + hour, + come, + doctor, when, chemo, + treatment, + bad, + start, + help, now, + give, + time, with, can	25	0.3623	Moderate effects
4	place, + cell, white, even, may, out, + make, + want, + know, + drug, can, up, + week, + would, + do, + count, + feel, + give, like, + have	2	0.02898	Reason for injection
5	+ week, + bone, + not, + have	1	0.01449	Bone pain

Table 6 Messages from Table 4 restricted to "Pain"

Cluster #	Description	Frequency	Percentage
1	+ injection, + agree, + side effect, severe, side, + hurt, but, + drug, + effect, only, down, + experience, chemo, + good, + round, + make, with, + do, + day, in	15	0.4634
2	bring, case, + week, up, + keep, same, + receive, back, during, much, + leg, little, blood, ever, + would, body, + count, + shot, + give, as	5	0.1219
3	+ eat, before, + call, walk, no, into, + hour, + come, + doctor, when, chemo, + treatment, + bad, + start, + help, now, + give, + time, with, can	16	0.3902
4	place, + cell, white, even, may, out, + make, + want, + know, + drug, can, up, + week, + would, + do, + count, + feel, + give, like, + have	0	0
5	+ week, + bone, + not, + have	1	0.02439

More than half of the original messages contain the word, pain. Of that number, most (85%) are in cluster 1, indicating severe pain, or in cluster 3, indicating moderate pain. We look at concept links to see how words are connected in the messages. These concept links are quite similar to the link graphs of market basket analysis and are computed in a similar fashion. Figure 19 gives the words connected to pain. It indicates that the pain can cause discomfort, or that it can be very severe; that is, pain as in cluster 1, or pain as in cluster 3.

Figure 20 gives an expanded link to see what is connected to both pain and discomfort.

Clinical Data Mining to Discover Optimal Treatment Patterns

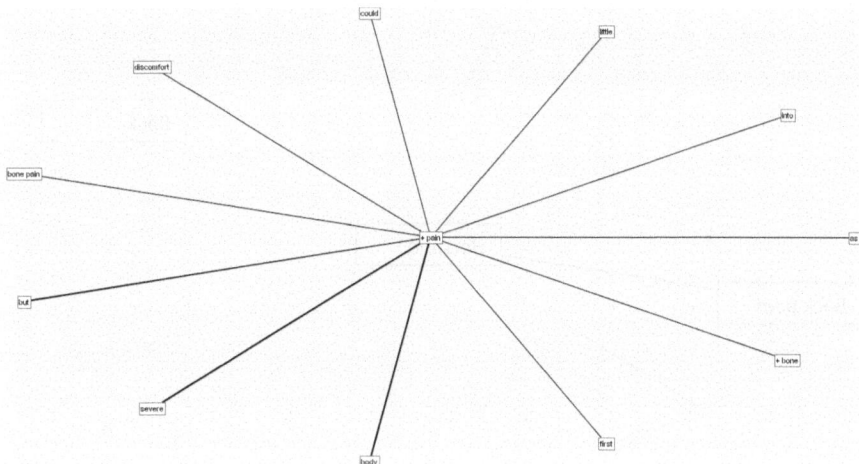

Fig. 19 Words connected to pain

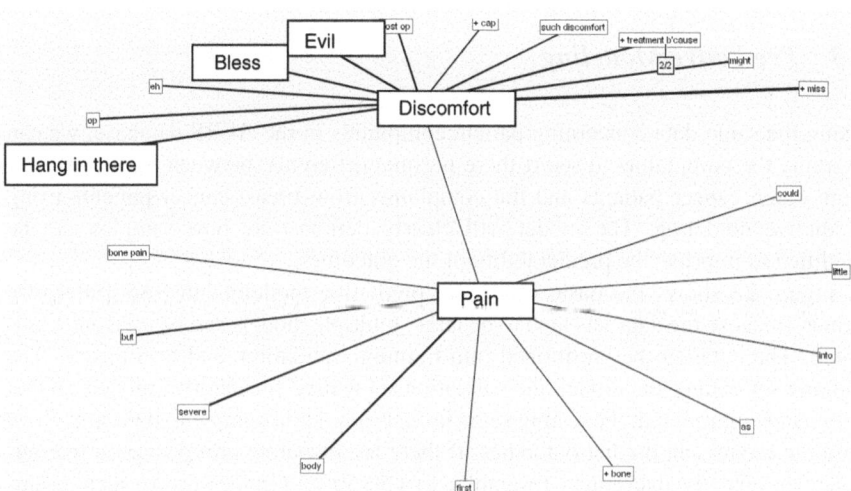

Fig. 20 Expanded links to pain and discomfort

Two of the terms relate to encouragement to continue in treatment, likely because the pain will stop once treatment with the drug is discontinued. Figure 21 examines links to the term, horrible, which indicates a severity that is greater than just discomfort, according to the sentiment of the patients. It indicates that horrible is connected to back and back pain, both of which show that the severity is in the back.

Once the text analysis is performed, we can look at work connections using these concept links. It is a visualization that is extremely useful.

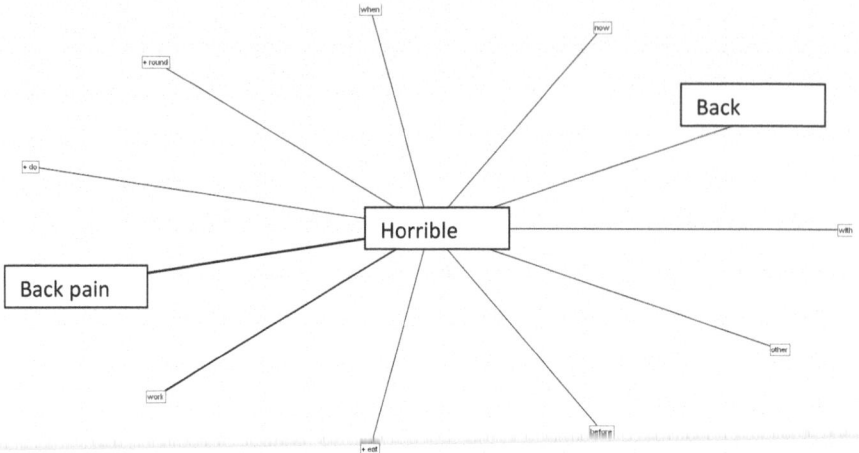

Fig. 21 Links to term, horrible

7.3 Predictive Modeling

Using the same data concerning patient complaints in the AERS database, we can examine the complaints to see if there is some difference between the complaints from colon cancer patients and the complaints from breast cancer patients using predictive modeling. The model will clearly demonstrate how samples can be modified to improve the predictability of the outcomes.

Figure 22 shows the basic outline of predictive modeling in SAS Enterprise Miner. Because the data sets tend to be large, multiple models can be used and compared. The data can be partitioned into training, validation, and testing sets. The training set defines the model; the validation set is used for iterative models so that they can be optimized; the testing set is used as a holdout sample to determine how well the model can predict outcomes. If there are disparate group sizes in the outcome variable, the data can be resampled to a 50/50 split in the groups; such resampling is a routine part of predictive modeling.

The three basic models are decision trees, neural network models, and regression. There are several different versions of these basic types available. The final step is to compare model predictions; we use misclassification for categorical outcomes and average error for continuous outcomes. There are additional information measures available that can be used to make comparisons as well.

Figure 23 shows receiver operating curves for the various models. This is a way to compare false positives versus false negatives for binary outcomes. The model that covers most of the area in the box is the optimal model. Note that the training data results look better compared to the validation and test results; for this reason, standard logistic regression will virtually always give inflated results that should be validated in some way. The straight line in the graphs indicates random chance; the

Clinical Data Mining to Discover Optimal Treatment Patterns

Fig. 22 Diagram of predictive modeling in SAS enterprise miner

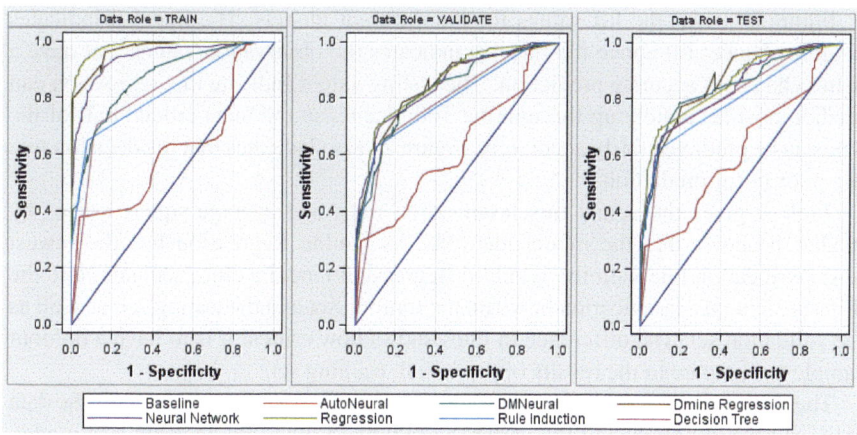

Fig. 23 Receiver operating curves to compared models

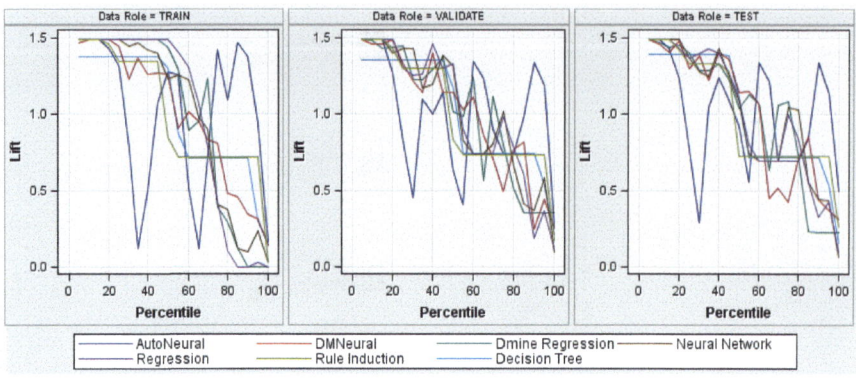

Fig. 24 Lift values by model

Table 7 Misclassification rates and average squared error by model

Model	Test: misclassification	Train: misclassification	Valid: misclassification	Train: average squared error	Valid: average squared error
Dmine regression	0.221	0.0814	0.220	0.063	0.196
Regression	0.230	0.0656	0.215	0.044	0.152
DMNeural	0.235	0.205	0.229	0.141	0.156
Neural	0.243	0.130	0.219	0.0973	0.150
Decision tree	0.258	0.241	0.275	0.146	0.163
Induction rule	0.283	0.263	0.284		
AutoNeural	0.307	0.292	0.297	0.255	0.293

model curve should be above this line or it has no predictive ability whatsoever. Figure 23 shows that regression is optimal.

Figure 24 gives the lift values for the different models. The line 1.0 indicates random chance; lift above the line 1.0 indicates the observation values that have a better chance of accurate prediction. The testing values indicate that regression can predict most accurately up through the 50th decile, so the most crucial half of the data can be predicted fairly accurately. Figure 24 also indicates that the decision tree is a poor fit for prediction.

Table 7 gives the actual misclassification rates and average squared error by model. It shows that the model identified as Dmine Regression has the lowest misclassification rate with the standard Regression model a close second. Note the difference in misclassification between the training set and the testing set, as well as the validation set. The difference in rates shows how critical it is to have a holdout sample to compare to the results of the initial, training set.

The decision tree, while a poor fit to the data, can be used to examine how the data split, and the importance of the variables. Figure 25 gives the decision tree model.

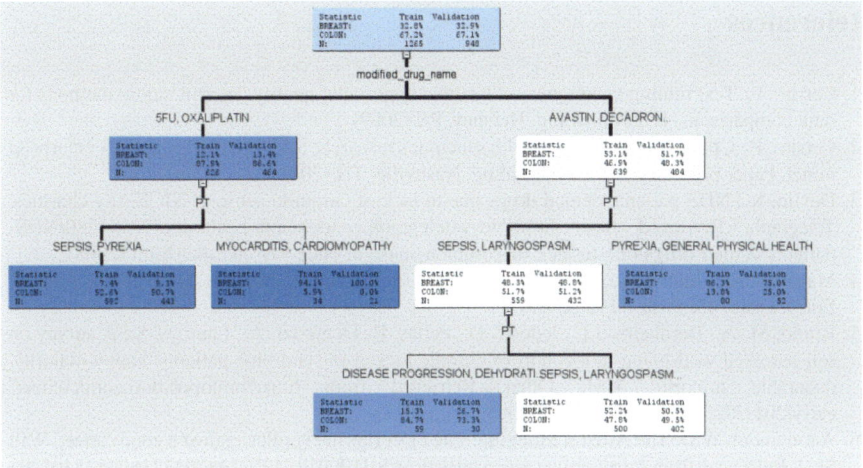

Fig. 25 Decision tree model for type of cancer

It indicates that the first split is based upon the drugs used. The second split is on the nature of the complaint, while the third split also includes disease progression.

To be more accurate, we will need to include patient diagnoses and patient outcomes in relation to the complaints. This analyses is ongoing.

8 Discussion

Large healthcare databases are now readily available for analytics investigations. They can find meaningful ways to optimize patient care and patient outcomes by applying data mining techniques, including text mining and market basket analysis as well as predictive modeling.

Healthcare data are difficult to investigate because of the diffuse nature of the outcomes, the definition of customer, the fact that the federal and state governments pay for almost 50% of all healthcare treatments, and the significant amount of government regulation. These problems will only become worse with the recently passed legislation with a mandate to purchase health insurance and the increased regulation of every aspect of healthcare. There is a continual analysis that the current rate of increase in costs for healthcare are unsustainable. In addition to examining optimal treatments for patient care, we need to investigate the issue of sustainability in healthcare. Otherwise, rationing will become a reality based upon definitions of quality of life.

Acknowledgment The author wishes to acknowledge the support of Dr. John Cerrito, PharmD, concerning the nature and effects of the drugs discussed in this chapter.

References

1. Cerrito, P.: Text mining techniques for healthcare provider quality determination: methods for rank comparisons. IGI Publishing, Hershey, PA (2009)
2. Cerrito, P.: Clinical trials versus health outcomes research: SAS/STAT versus SAS enterprise miner. Paper presented at the PharmaSug, Nashville, TN (2011)
3. Devlin, K.: NHS patients denied drugs due to lack of common sense at NICE, say charities. Telegraph. Retrieved from http://www.telegraph.co.uk/health/healthnews/3531280/NHS-patients-denied-drugs-due-to-lack-of-common-sense-at-Nice-say-charities.html (2008)
4. Mason, A.R., Drummond, M.F.: Public funding of new cancer drugs: is NICE getting nastier? Eur. J. Canc. **45**, 1188–1192 (2009)
5. Bruno, M.-A., Bernheim, J.L., ledoux, D., Pellas, F., Demertzi, A., Laureys, S.: A survey on self-assessed well-being in a cohort of chronic locked-in syndrome patients: happy majority, miserable minority. BMJ Open. Retrieved from http://bmjopen.bmj.com/content/early/2011/02/16/bmjopen-2010-000039.full (2011)
6. Anonymous-WSJ, The Avastin Mugging: The FDA rigs the verdict against a good cancer. Wall St. J. Retrieved from http://online.wsj.com/article/SB10001424052748704271804575405203894857436.html (2010)
7. Perrone, M.: FDA delays decision on breast cancer drug Avastin. AP Associated Press. Retrieved from http://www.msnbc.msn.com/id/39239537/ns/health-cancer/ (2010)
8. Anonymous-maintenance study. Bevacizumab and combination chemotherapy in treating patients with previously untreated metastatic colorectal cancer that cannot be removed by surgery. (NCt00797485). Retrieved from http://clinicaltrials.gov/ct2/show/NCT00797485 (2009)
9. Anonymous-bevacizumab. Bevacizumab and cetuximab for the treatment of metastatic colorectal cancer. National Health Service, London (2009)
10. Giuliani, F., Vita, F.D., Colucci, G., Pisconti, S.: Maintenance therapy in colon cancer. Canc. Treat. Rev. **26**(Suppl 3), S42–45 (2010)
11. Hay, J.W.: Using pharmacoeconomics to value pharmacotherapy. Clin. Pharmacol. Therapeut. **84**(2), 197–200 (2008)
12. Puthillath, A., Patel, A., Fakih, M.G.: Targeted therapies in the management of colorectal carcinoma: role of bevacizumab. Onco. Targets Ther. **2**, 1–15 (2009)
13. Castells, M.C., Tennant, N.M., Sloane, D.E., Hsu, F.I., Barrett, N.A., Hong, D.I., et al.: Hypersensitivity reactions to chemotherapy: outcomes and safety of rapid desensitization in 413 cases. J. Allergy Clin. Immunol. **122**, 574–580 (2008)
14. Togashi, Y., Kim, Y.H., Masago, K., Tamai, K., Sakamori, Y., Mio, T., et al.: Pulmonary embolism due to internal jugular vein thrombosis in a patient with non-small cell lung cancer receiving bevacizumab Retrieved from http://www.springerlink.com/content/0208402248n7mn4k/ (2010)

Exploring the Parallels Between a Hospital Pharmacy and a Distribution Center

Jennifer A. Pazour and Russell D. Meller

1 Introduction

The pharmaceutical supply chain is the means through which prescription medications are manufactured, transported, stored, and delivered to patients. Pharmaceutical distributors provide the link between manufacturers, which produce medications, and pharmacies, which administer medications to patients. By storing and managing inventory, distributors are responsible for ensuring that pharmacies are equipped with the needed medications to care for patients.

Because pharmaceutical distributors handle 85% of all prescription drug sales in the USA [7], pharmaceutical distributors play a vital role in our economy and impact the cost and quality of our healthcare system. One example of a pharmaceutical distributor is a wholesale distributor, such as Morris & Dickson or Cardinal Health. Another example is a company-owned distribution center, like a Walgreens or Wal-mart pharmacy distribution center. Typically, company-owned distributors handle only a small percentage of a company's total stock keeping units (SKUs), relying on wholesale distributors for the remaining products [5].

Two types of pharmacies are hospital and retail pharmacies. A hospital pharmacy resides within an acute care facility and administers medications to patients during their stay at the hospital. Medications are delivered and stored in a pharmacy, and then when demand warrants it, are transported from the pharmacy to a clinical floor for a nurse to administer, typically in unit-dose packages. A retail pharmacy fills outpatient prescription requests and is responsible for the safe storage and dispensing of medications to patients.

J.A. Pazour (✉) • R.D. Meller
Department of Industrial Engineering, University of Arkansas,
Fayetteville, AR 72701, USA
e-mail: jpazour@uark.edu; rmeller@uark.edu

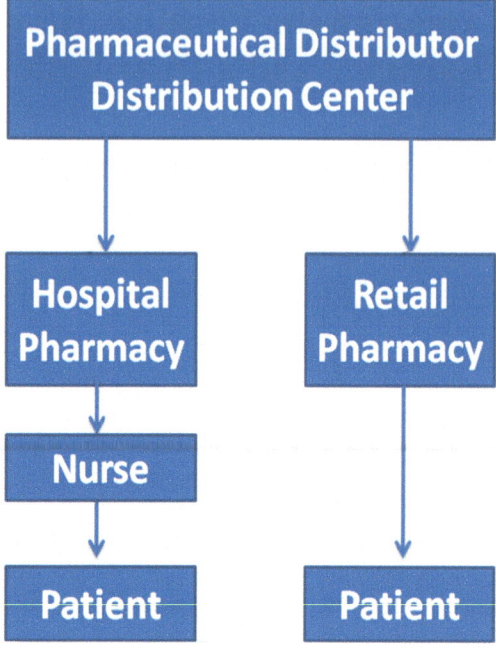

Fig. 1 The pharmaceutical supply chain from the distributor to the patient

In our research we focus on the pharmaceutical supply chain from a distributor to a patient as represented in Fig. 1. We discuss two types of distribution centers within the pharmaceutical supply chain: a traditional distribution center and an on-site distribution center, the hospital pharmacy.

1.1 A Traditional Distribution Center

A traditional distribution center stores a set of SKUs to be distributed to an entity downstream in the supply chain. The basic operations in a distribution center are (a) to receive products from suppliers, (b) to store the products until requested from customers, (c) to fulfill the orders by retrieving the needed products and assembling the orders, and (d) to transport the orders.

Distribution centers provide many benefits. Distribution centers store inventory to balance supply with demand, allowing for quicker response to customer demand. The presence of a distribution center can reduce transportation costs by consolidating shipments. Distributor centers allow manufacturers to efficiently produce large quantities of products and through break-bulk activities, ship customers small quantities of a variety of products. Also, value-added processes are often conducted at a distribution center.

Gu et al. [8] provide a recent review of academic literature associated with distribution center operations and classify the literature within a unifying framework based on distribution center functions. A companion paper [9], reviews the academic literature in the area of distribution center design, performance evaluation, practical case studies, and computational support tools.

In the next section we provide further details on distribution centers used in the pharmaceutical sector.

1.1.1 A Pharmaceutical Distribution Center

Manufacturers outsource the distribution of much of their products to distributors, rather than engage in director customer delivery [5]. Consequently, pharmaceutical distributors are responsible for ensuring that hospital and retail pharmacies are supplied with the needed medication to care for patients. The pharmaceutical distribution business is characterized by several dynamics.

First, pharmaceutical distribution is characterized by high-valued products that have an expiration date. These characteristics motivate pharmacies to minimize their on-hand inventories by requesting frequent orders, leading to medication shipments being received daily or even multiple times a day. In 1998, 67% of distributors made deliveries five times a week; another 31% made deliveries six times a week [5].

Second, the number of delivery sites from which pharmaceutical products are dispensed is large. In 1998, pharmaceuticals were dispensed from 135,000 sites in the USA including 20,000 independent pharmacies, 19,000 chain pharmacy outlets, 60,000 clinics, 7,000 hospitals, and 12,000 mass merchandisers and food stores [5]. The median number of shipping points served per distributor in 1998 was 525 [5]. Therefore, pharmaceutical distributors mediate pharmacy's desire for small quantities of a diverse number of products with manufacturers that produce large quantities of a few products.

Third, pharmacies place stringent requirements on order lead-time, order accuracy, and availability of a diverse product line. Extremely short lead time requirements of an hour or less can be placed on the order-fulfillment process [11]. Furthermore, it is common for pharmaceutical distribution facilities to consider more than 20,000 active SKUs [3].

Pharmaceutical distributors have utilized automation to respond to customer needs by driving greater volumes through distribution centers at faster speeds. In particular, order-fulfillment technology allows pharmaceutical distributors better response to their customers' needs by processing orders more quickly and accurately. One type of customer for a pharmaceutical distribution is a hospital pharmacy.

1.2 A Hospital Pharmacy

Hospital pharmacies oversee the procurement of medications used within the hospital. Because hospital pharmacies store a supply of medications and then are tasked with delivering medications in unit-dose packages to the clinical floor for the nurse to administer, we consider them as on-site distribution centers. The flow of medications through the hospital can be categorized into the following functional processes: ordering, receiving, storage, order-fulfillment, and transportation.

Some pharmacies have automated the ordering process by investing in inventory technology that tracks inventory and par levels. In these systems, a pharmacy technician reviews the order, making necessary changes where applicable. The order is then transmitted to a supplier or a pharmaceutical distributor, where it is fulfilled. The order is transported from the distribution facility to the hospital pharmacy, typically through less-than-truckload or parcel carriers.

Upon arrival of the physical order to the pharmacy, a pharmacist or pharmacy technician receives the product and validates that the correct number of packages have been received. The medications are stored in various storage locations within the pharmacy. These can include carousel systems, refrigerated storage, narcotic vaults, open shelving, bins, and repackaging technologies. If medications are not received in unit-dose packages, repackaging processes occur.

Once medications are in unit-dose form, there are two main strategies for unit-dose medication distribution: cart-less and cart-fill. In both systems, unit-dose medications are stored on the hospital floor in a medication cart. In a cart-less system, available medications are sorted by medication types, whereas in a cart-fill system the medications are sorted by patients. In both systems, order-fulfillment is conducted by the central pharmacy; the only difference is in a cart-less system medications are picked and stored by medication-type, while in a cart-fill system medications are picked and stored by patient.

Medications on clinical units are typically stored in a controlled environment, often times in automated dispensing cabinets. Typically, a database manages information about the inventory in each automated dispensing cabinet, such that when a storage location reaches its par level, an order is generated and sent to the pharmacy. The orders generated are aggregated into a pick list that pharmacy technicians fill and pharmacists verify. The medication order is transported to the clinical unit and stored in the proper location until retrieved by a nurse for medication administration to a patient. Numerous safety and quality checks are implemented such that a nurse administers the correct medication for the correct patient at the correct time in the correct dose.

Controlled substances have additional regulations from the government, requiring additional monitoring and controls to be in place. From the time a controlled substance is delivered to the hospital until it is administered to the patient, the medication must be tracked by the hospital staff, which includes a number of tasks such as checking the ordered medication against the patient's medical record, documenting for administrative records, and reconciling control substance records after each shift [13].

Hospital administrators and pharmacists have to manage a very complicated distribution network without the proper training or educational backgrounds to do so efficiently [13]. Consequently, one of the goals of our research is to compare hospital pharmacies to traditional distribution centers. From this comparison, best practices from distribution center design can be identified and transferred to hospital pharmacies, where applicable. In the next section we make an initial comparison.

2 Comparison Between a Traditional Distribution Center and a Hospital Pharmacy

We view a hospital pharmacy as an on-site distribution center as both perform similar functions: ordering, receiving, storage, order-fulfillment, and transportation. Viewing the hospital pharmacy as such, we make the following comparisons between a traditional distribution center and a hospital pharmacy.

- *Ordering and Receiving.* Distribution centers order and receive products so that transportation can be consolidated and inventory costs can be reduced at retail locations. Similarly, a hospital pharmacy orders and receives medications so that individual clinical units can experience reduced transportation and inventory costs.
- *Storage Process.* Distribution centers store inventory to be able to respond to customer requests. Likewise, hospital pharmacies hold stocks of medications in order to be able to respond to patient prescriptions and emergency situations. Due to the difficult-to-predict demand for medications and the emergency environment, higher levels of safety stock are typical in hospital pharmacies.
- *Order-Fulfillment Process.* Distribution centers experience demand for products that must be picked and transported. Typically, distribution centers receive products in large quantities from manufacturers and provide retail locations with small quantities of a variety of products through the order-fulfillment process. Orders can be fulfilled at different levels, ranging from piece-level, carton-based, or unit-load fulfillment. A hospital pharmacy also experiences demand for medications from clinical units that must be picked and transported. Medications typically arrive to the hospital pharmacy at the piece or case levels and are supplied to the clinical units in unit-dose form. Therefore, delivering unit-dose medications to the hospital floor can be viewed as piece-level order-fulfillment.
- *Multichannel Fulfillment System.* Distribution centers typically have a multichannel fulfillment system, especially when the storage and picking activities are separated. A separate picking area, called a fast-pick area, is a subregion of the warehouse that concentrates picking within a small physical space to reduce pick costs and increase responsiveness to customer demand. The fast-pick area often utilizes order-fulfillment technology such as pick-to-light, A-Frame systems, carousel systems, or picking machines. In a pharmacy, medications can be picked from traditional bins, carousel systems, or fully-automated repackaging systems. An additional channel devoted to controlled substances and medications formuwlated for patient-specific needs also exists.
- *Labor and Infrastructure Trade-off.* In a distribution center, labor is the dominant expense. A trade-off lies in balancing the capital expenditures in infrastructure with the labor savings and increased quality gained by the technology. Hospital pharmacies are also labor intensive and investments in infrastructure can also aid in reducing labor and increasing quality. Even though some of the specific technology varies between a distribution center and hospital pharmacy, there are also similarities; for example, carousel systems are common in both.

- *Value-Added Processes.* Distribution centers can provide value-added processes, such as delayed product differentiation and packaging improvements. In a pharmacy, repackaging medications from bulk quantities to unit-dose packages is a value-added process used to increase patient safety through barcode-enabled point-of-care (BPOC). Other value-added processes in a hospital pharmacy include validating that a patient is not allergic to the prescribed medication and creating dilutions.
- *Facility Variety.* Distribution centers have varying demand, infrastructure, product, and customer profiles. Likewise, each pharmacy has unique characteristics. Some examples include the lead time allowed from when the doctor prescribes the order to when the medications must be on the hospital floor, the volume of medications, the diversity of medications, and the type of software and infrastructure used for distribution.

In the next section we provide an overview of the technologies used in a traditional distribution center and a hospital pharmacy, with special emphasis on the order-fulfillment process.

3 Order-Fulfillment Technologies in Pharmaceutical Distribution

In both a traditional distribution center and a hospital pharmacy, order-fulfillment, the process of retrieving products from storage in response to a specific customer request, is one of the most critical and expensive tasks in the distribution process because of its simultaneous impact on the cost, quality, and accuracy of the process. Therefore, specifying an effective order-fulfillment process is an important aspect in distribution center and hospital pharmacy design. Our research will focus on the piece-level fulfillment process as it is the most common fulfillment level used in pharmaceutical distribution.

When item-level demand is high, the piece-level order-fulfillment process can be very labor-intensive. Furthermore, when labor wages are also high, this type of order-fulfillment becomes a good candidate for automation. And in the pharmaceutical industry, an added motivation for automation is the requirement of extremely high-order accuracy and short lead time requirements.

3.1 Distribution Center Order-Fulfillment Technologies

There are three typical piece-level order-fulfillment strategies in a distribution center—picker-to-stock, stock-to-picker, and an automated dispensing system—with various technologies used to automate the order-fulfillment process associated with each strategy (see Table 1).

Table 1 Piece-level order-fulfillment technology

Picker-to-stock	Stock-to-picker	Automated dispensing
Pick-to-light	Carousels	A-Frame systems
Sortation systems	Vertical lift modules	Dispenser systems
Conveyor systems	Mini-load AS/RS	
Pick-and-pass systems	Picking machines	
	Put-to-light systems	

With a picker-to-stock strategy, an operator visits fixed locations to make a pick. Picking is often performed in a consolidated area, known as a fast-pick area. When warranted due to high labor wages or the need for high picking accuracy, a pick-to-light system is employed in order to reduce the amount of time an operator spends searching for the next pick and/or for increasing accuracy. To decrease walking, orders may be batched picked such that cases are picked but less-than-case quantities are shipped. To facilitate this strategy, a split-case sortation function is performed to sort the correct items into the correct order. Additionally, conveyor systems may be used to transport items, further reducing the amount of walking performed. In a pick-and-pass system, each pick station has buffer spaces such that the main flow of totes does not become blocked by totes waiting for picking.

With a stock-to-picker strategy, materials to be picked are transported to the operator. Technologies to facilitate this strategy include carousels, vertical lift modules (VLMs), mini-load automated storage and retrieval systems (AS/RSs), picking machines, and put-to-light systems.

Finally, with an automated dispensing strategy, automated order picking machines can be used to completely eliminate manual picking; common automated dispensing technologies are the A-Frame and dispenser systems. However, even in automated dispensing systems, operators typically perform manual replenishment.

3.2 Hospital Pharmacy Order-Fulfillment Technologies

Order-fulfillment technology is used in a hospital pharmacy to increase patient safety by reducing picking errors, as well as to reduce costs through labor, inventory, and space savings. Some example technologies include vertical carousel systems, repackaging technologies, and automated dispensing cabinets. Vertical carousel systems store medications and can reduce the labor associated with the order-fulfillment process by bringing the medication closer to the technician. Repackaging technologies package bulk medications in unit-dose form. Automated dispensing cabinets are medication storage devices on the clinical units and are commonly what the pharmacy technicians are fulfilling orders to.

These technologies, as well as order-fulfillment technologies used at distribution centers will be the focus of the next section. We provide a description of the order-fulfillment technologies and review the analytical models associated with these technologies.

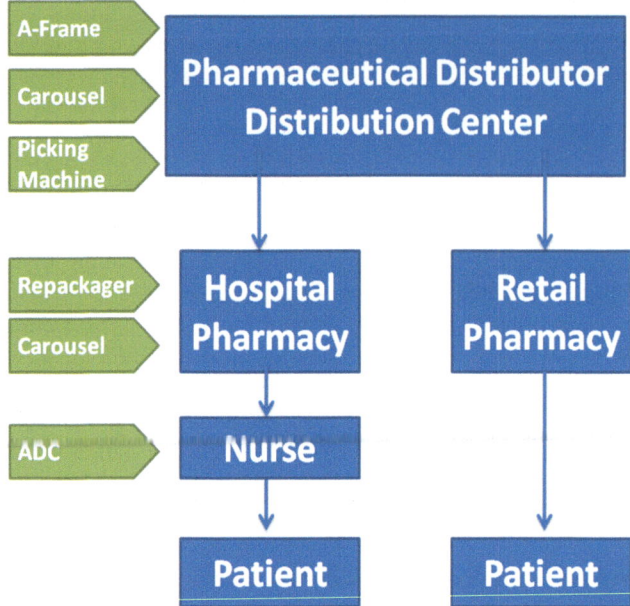

Fig. 2 Where order-fulfillment technology is applied in the pharmaceutical supply chain

4 Modeling of Order-Fulfillment Technologies in Pharmaceutical Distribution

We focus on models that analyze order-fulfillment technology in pharmaceutical distribution. Unlike the consumer-product supply chain, the healthcare supply chain has not received much attention from the academic-research community and thus an opportunity for impact in this line of study exists.

We have identified five technologies used in the order-fulfillment process as the main focus of our discussion: A-Frame systems, carousel systems, picking machines, automated dispensing cabinets, and unit-dose repackagers. The former three technologies focus on pharmaceutical distribution from a traditional distribution center to a hospital pharmacy, while the latter two technologies focus on order-fulfillment within a hospital. Figure 2 illustrates where each identified technology is applied in the pharmaceutical supply chain.

In addition to discussing each of the identified technologies, we review the models available in the academic literature to analyze piece-level order-fulfillment technologies prevalent in pharmaceutical distribution.

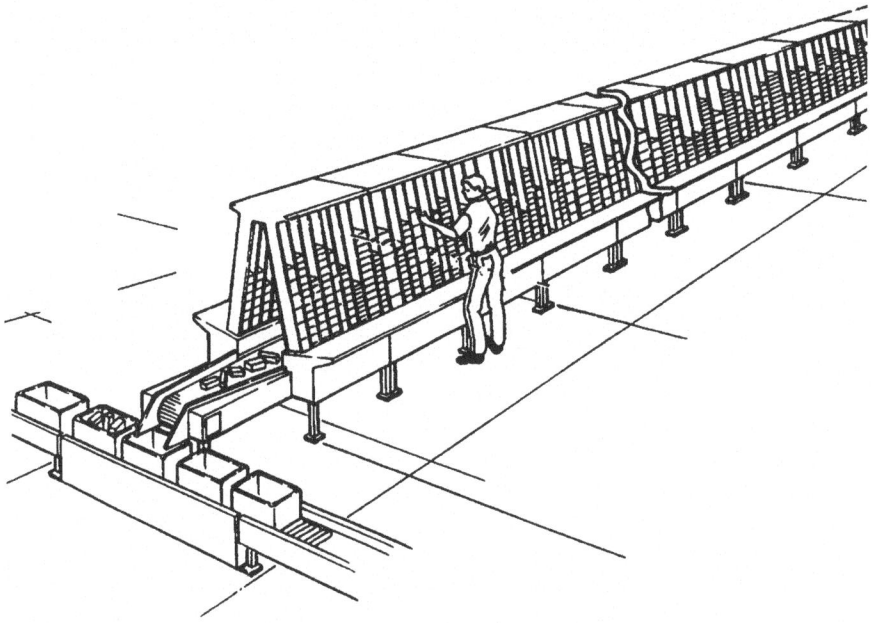

Fig. 3 Depiction of an A-Frame system (Figure Courtesy of MHIA)

4.1 A-Frame Systems

A common automated dispensing system used in pharmaceutical distribution is the A-Frame system. As illustrated in Fig. 3, the products are placed in side-by-side channels of varying magazine length on two sides of a collection belt, forming an "A" and the items are automatically dispensed onto the belt that passes inside the tunnel created by the frame. Orders are filled one at a time, with products then directly fed into customer order totes and transported via a conveyor to another picking station or a packing station. While the machine is operating, manual replenishment activities of the product channels can be performed with no impact on the dispensing operations.

A-Frame systems are capable of filling orders quickly (up to 750,000 picks/day, 1,200–2,400 orders per hour) with incredible order accuracy (>99.95%) [6]. They are common in pharmaceutical, electronic, and cosmetic distribution facilities, where large volumes of highly valued items are fulfilled in small order sizes with stringent constraints on lead-time and order accuracy.

The main decisions to be addressed when designing an A-Frame system are determining which medications (and in what quantities) should be stored in the A-Frame, as there tends to be capacity constraints due to the infrastructure costs related to the A-Frame system. In addition, A-Frame systems are complex machines with interacting components that effect throughput. There are three primary criteria used to determine which SKUs to assign to the A-Frame, and in what quantities.

- Replenishment
- Economical Trade-offs of Infrastructure and Picking Costs
- Throughput

The first criterion, replenishment, has been researched by Bartholdi et al. [2], Caputo and Pelagagge [6], and Jernigan [12]. Caputo and Pelagagge [6] design a decision-support tool for modifying the system setup once demand is observed for a period of time. Through a rule-based heuristic, the system setup can be tuned for the next period, evaluating product substitutions, modifying the number of channels, and determining the reorder level and the maximum quantity to be dispensed. A significant component of their decision-support tool is the consideration of channel fill levels. The authors' decision support tool suggests through empirical testing that these levels tended to be much higher in practice than appropriate.

A method for the assignment and allocation of SKUs to an A-Frame based on the number of replenishments and pick costs is presented by Bartholdi et al. [2] and Jernigan [12]. They do not consider infrastructure costs in their model, but instead assume that each order-fulfillment area has a known storage capacity. Moreover, they consider minimizing the number of restocks to the A-Frame and assume that each replenishment, regardless of the number of items, has a fixed replenishment time and can occur instantaneously (i.e., no allowance for safety stock or stochastic demand). Their replenishment schedule allots space in terms of a fraction of the available storage capacity, which does not address the discreteness of channels, the different lengths of channels, or if an SKU can fit or fill the space allotted to it. They extend their models to consider several time periods and reassignment costs.

Liu et al. [14] extend Jernigan's work by analyzing a system with both an A-Frame and an alternate horizontal-dispenser (HD) system. In an HD system, the channels are arranged horizontally versus vertically and are capable of ejecting more than one item at a time. They incorporate the discreteness of channels, but assume that all SKUs have the same physical dimensions and ignore pick costs, infrastructure costs, and throughput. They determine the number of channels to provide to each SKU based on minimizing the number of restocks required, assuming restock costs are not a function of the number of items replenished. They assign SKUs to either the A-Frame or HD system using a greedy heuristic based on a labor-efficiency ranking.

Yaohua and Yigong [30] analyze a parallel-dispenser system, which is also an alternative automated dispensing technology. A parallel-dispenser consists of three kinds of channels (i.e., launching, pushing, and replenishing channels) and can eject more than one item at a time. Unlike an A-Frame system, the sequence of orders does impact throughput in a parallel-dispenser system. Therefore, they develop a heuristic for sequencing orders to minimize the total picking time. They assume the allocation and assignment of SKUs to the parallel-dispenser system is given and do not consider infrastructure costs of the system.

Reference [22] analyze the design of an A-Frame system in a distribution center, which includes evaluating the technology to arrive at the best-cost solution that meets throughput considerations. Their research addresses the question of which SKUs to assign to an A-Frame system, and in what quantity, considering labor, infrastructure, and throughput. This is a critical decision as it has a direct impact on the efficiency of an A-Frame system through balancing replenishment costs, picking costs, and infrastructure costs, as well as the throughput of the process.

A math programming-based approach to determine the amount of A-Frame infrastructure investment, as well as the assignment and allocation of SKUs to the A-Frame, is provided in [22]. They allocate channels in an A-Frame system based on an inventory modeling approach. Next, they address throughput considerations explicitly by developing analytical models for the throughput of an A-Frame and heuristics to adjust the allocation and assignment of SKUs in order for the A-Frame to meet a throughput constraint. They prove that the best-case allocation for labor savings is the worst-case allocation for A-Frame throughput. They test their methodology on a pharmaceutical industry example and determine the impact of parameters on their model.

Their methodology indicates that A-Frame systems provide the greatest impact on labor savings when a distribution center has high item commonality, small order sizes, and high skewness levels and on throughput when a distribution center has low item commonality, small order sizes, and low skewness levels.

4.2 Carousel Systems

A carousel system is a stock-to-picker piece-level technology that consists of storage locations that rotate around a closed loop. The carousel system is served by an operator (either human or robotic) that is at a fixed-picking position and the storage locations rotate in either direction to bring items to an operator.

Carousel systems are used in both a traditional distribution center and a hospital pharmacy and have average picking rates of 200–400 lines per person hour [27].

In a distribution center, carousels are commonly utilized in pods with more than one carousel unit per pod, as shown in Fig. 4. This configuration allows one unit in the pod to retrieve the next pick, while the operator is retrieving an item from another unit in the pod. A carousel system rotates such that the requested items are in front of the operator. Pick lights inform the operator of the position and quantity of the product to be picked. After picking the items and putting them in the correct tote, the operator walks to another carousel in the pod and picks again. The rotation time of a carousel is a function of the length of a carousel; consequently, the number of carousels in a pod can be varied to ensure that the carousels are not the bottleneck in the system, constraining the human order picker. As carousel systems

Fig. 4 Multiple carousel pods

are typically throughput constrained (rather than space constrained) [17], multiple carousel pods may be implemented to meet throughput requirements. One operator is assigned to each pod of carousels and batch picking is usually performed.

The literature on carousel systems used in a distribution center is abundant and can be classified into the following problem areas: rotation strategies, performance models, and assignment of storage locations. For a recent review, see Hassini [10] and the references within. The majority of the literature focuses on models for single carousel systems picking a single item.

Hospital pharmacies that use carousel systems typically pick a batch of medications and often have multiple carousels. Multiple medications are stored in a single bin to increase storage densities. Whereas specific models for carousels used in a

Fig. 5 Depiction of a picking machine

hospital pharmacy have not received attention from the academic community, Meller and Klote [17] could be applied to determine the throughput of a carousel with human order-pickers picking batches of orders.

In the next section models for an additional stock-to-picker system, a picking machine, are presented.

4.3 Picking Machines

Picking machines, also known as automated-storage-and-order-fulfillment systems, are an example of an alternate stock-to-picker piece-level fulfillment technology. Picking machines incorporate pick-to-light technology, put-to-light technology, conveyor systems, and carousels or mini-load AS/RSs in order to increase throughput and provide high product density. Depending on the system configuration, a picking rate of up to 1,000 order lines per person-hour is possible [28], which indicates why this technology is seen in industries that have a large number of active SKUs and small order sizes. These characteristics, as well as the security of the items and the ability to conduct lot-tracking, make picking machines common in the pharmaceutical industry.

A picking machine consists of numerous pick stations and a system of carousel units or a mini-load AS/RS to provide storage. An integrated conveyor system transports the requested totes to and from the storage area and the picking stations, as depicted in Fig. 5.

Therefore, even though there are numerous pick stations, each pick station utilizes the same carousel system or mini-load AS/RS for storage. The required orders' shipping containers are transported via a conveyor to a pick station. At the same time, the requested SKUs are retrieved automatically from the storage system and are also sent to the pick station. Displays indicate to the operator both the position and the quantity of the product to be picked, as well as the position of the container to which the SKU should be transferred (the put operation). The replenishment of the storage area can be conducted concurrently with the order-picking process or handled off-shift.

Two conference papers with preliminary simulation results address picking machines. Perry et al. [25] use a discrete-event simulation model to assist in the physical system design of a picking machine. A simple, expected-value model is used for initial design variable values, which are then modified based on throughput requirements. Their testing indicates that the conveyor system is the bottleneck on system throughput. Raghunath et al. [26] describe an interactive and flexible simulation structure for a picking machine.

Picking machines have been mentioned in the literature as future research by Bozer and White [4]: "A possibility is to investigate the use of remote picking stations where each station is interfaced to the storage/retrieval system via a closed conveyor loop. Such a system allows each picking station access to the aisles." Also, Park et al. [19] classify mini-load AS/RSs into three categories, one mentioned as future research is a closed-loop conveyor: "mini-load systems containing a closed-loop conveyor, often called the remote order picking system, have a closed-loop conveyor system to deliver the containers that interconnects each aisle of the mini-load system with the remote order picking stations."

Because picking machines are touted as a lower-inventory alternative to carousel systems, Pazour et al. [21] develop a probabilistic model capable of quantifying the inventory differences between these two technologies. They also determine the throughput of a picking machine by analyzing its subsystems, the carousel storage system, the closed-loop conveyor, and the pick stations, independently. An expected cycle-time model is developed to determine the throughput of a carousel system with a storage and retrieval machine performing batch retrievals. A stability-condition model is applied to determine if the conveyor system will be stable for an expected throughput requirement. A case study comparing a picking machine to a carousel-pod system is presented to illustrate how a manager could use their analytical models to answer system design questions.

In the next section we focus on order-fulfillment technology used within a hospital.

4.4 Unit-Dose Repackaging Technologies

In hospitals, medication errors can occur during every step of the process, but occur most frequently during the prescribing and administering stages. In fact, the following startling statistic is given in a report entitled *Preventing Medication Errors* [1], "when

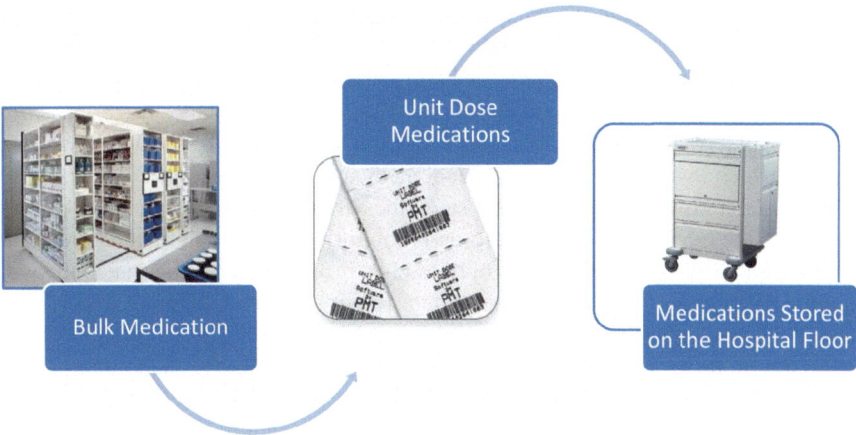

Fig. 6 Flow of medications through a hospital (source: Spacesaving.com, RnInsider.com, and RxShelving.com)

all types of errors are taken into account, a hospital patient can expect, on average, to be subjected to more than one medication error each day." One way to prevent such medication errors and increase patient safety is to administer medications in unit-dose packages [15]. To further increase patient safety, many hospitals today are implementing barcode-enabled point-of-care (BPOC) systems. These systems require that medications are in barcoded unit-dose packages and also that patients wear a barcoded bracelet. When a nurse administers the patient's medications, he or she must first scan the barcodes on the medication and the patient. These systems help ensure that the right medications reach the right patient at the right time by allowing barcodes on a patient's ID wristband and medication packaging to be checked against a database containing medication identification and physicians' orders.

Unfortunately, not all medications are available in unit-dose form. A 2008 survey of hospital pharmacists found that on average, hospitals could obtain only 56% of their total formulary in unit-dose form directly from the manufacturer [16]. Currently, if a hospital wants to use unit-dose dispensing, the choice is to either purchase prepackaged unit-dose medications directly from the supplier, utilize a third-party repackager, or purchase bulk supplies of medications from manufacturers and repackage them on-site into unit-dose packages [18].

Various levels of automation are currently being utilized to handle unit-dose medication repackaging on-site. As illustrated in Fig. 6, a typical process is to buy bulk medications from the supplier, package medications in unit-doses, and then store the unit-dose medications on the hospital floor, typically in automated dispensing cabinets.

Technology for repackaging medications in unit-dose form is classified into three levels: manual, semi-automated, and fully-automated. In a manual system, a technician retrieves the appropriate bulk medication and then has the laborious task of physically repackaging the bulk medication into unit-dose packages. A manual process typically uses little more than a device to seal the packaging with manual labor used for the repackaging and relabeling process.

The semiautomated option is typified by what is referred to as a *table-top unit*, consisting of a rotary table, an automated sealer and labeler, and a manual induction process, which involves placing individual medications into openings on the top of the rotary table. In a semiautomated system, the actual repackaging process is automated, but the retrieval of medications and monitoring processes are not. A technician is required to retrieve the appropriate bulk medication, fill the machine, and monitor the process. In addition, the system requires manual cleaning in order to prevent the cross contamination of medications.

Finally, a fully-automated system, also known as an automated repackager, removes a large portion of the labor associated with repackaging. A fully-automated system consists of a set of canisters, where each canister holds one medication. The medications travel down a common chute to the packaging and labeling units. Manual bulk replenishment of the machine is required. Because of the common chute, medications that would contaminate other medications cannot utilize an automated repackager. Therefore, these machines are not always able to hold all of the unit-dose medications on the formulary and some other system (i.e., a manual or semi-manual system) is still required.

Reference [23] develop a mathematical model that simultaneously determines which level of technology is warranted and how each medication that is not delivered to the pharmacy in unit-dose form should be repackaged subject to multiple constraints. This model has been integrated into a free Excel-based tool available to pharmacy directors. They test their model with data based on small, medium, and large hospitals and conduct sensitivity analyses to gain further insight. They illustrate how the results from their model can aid in incorporating qualitative aspects into technology selection. Their results show that a semiautomated repackaging system is the most economical technology alternative for most hospital pharmacy in-house repackaging operations. This result, however, is sensitive to quantitative factors like the number of unit-dose medications to repackage and the available labor, and to qualitative factors like, repackaging quality, pharmacist retention, and storage space constraints.

Once medications are in unit-dose form, the medications are stored on clinical units in automated dispensing cabinets, which are the focus of the next section.

4.5 Automated Dispensing Cabinets

Automated dispensing cabinets (ADCs) are medication storage devices that allow medications to be stored and dispensed near the point of care, creating a more decentralized medication distribution system with shorter nurse response times. An example of an ADC is a Pyxis machine shown in Fig. 7. Currently, 83% of USA hospitals use ADCs [24] and it is common to have one or more ADCs in each clinical unit.

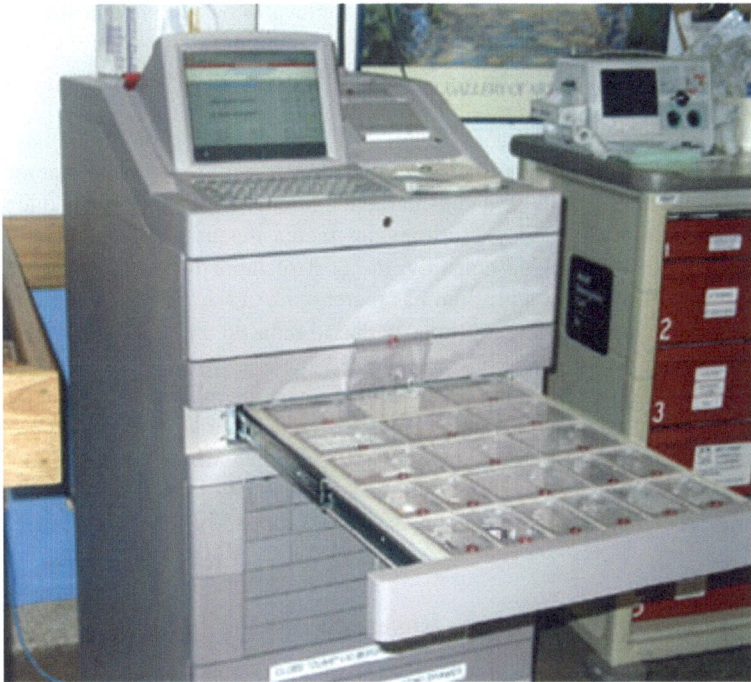

Fig. 7 An automated dispensing cabinet

Medications are stored within ADCs such that each medication has its own separate and segregated location. In the case of a matrix drawer, all medications within a drawer are accessible to a nurse and 48.5% of hospitals surveyed use matrix drawer configurations as the predominant configuration in their ADCs [24]. Typically, no scanning of medications is performed when a nurse removes the prescribed medications for a patient.

To reduce opportunities for wrong medication selection and dosing errors, medications should be located in the compartments such that similar medications are not in close proximity to one another. Medication similarity can be defined in terms of a medication's nomenclature, packaging, dosage form, risk level, and dosage concentration. A safe practice is to store different concentrations or dosages of a medication in a manner that will minimize the risk of accessing the wrong strength of the medication. The Institute of Safe Medication Practices (ISMP) has defined a list of medications that are look- and sound-alike medications and recommend that they should not be located near each other within the ADC. For example, it is best not to place metformin near metronidazole, as both are oral solid medications that sound-a-like but have very different uses (metformin is used to improve glycemic control in adults, while metronidazole is used to treat bacterial infections).

Due to capacity constraints on the ADC and medication expiration dates, the assignment and removal of medications from compartment locations within an

ADC occurs daily. Medication selection and removal is primarily determined based on the collective needs of the patients on a clinical wing. Determining the locations of medications in the ADC is typically a manual process performed by a pharmacist technician.

Wise et al. [29] conduct a cost-benefit analysis on the decision of purchasing an automated medication distribution system in a hospital unit. They cite that medication and distribution compose a considerable amount of economic costs due to the current medical practice of prescribing an increasing number and diversity of medications to patients. Consequently, the cost of labor to store, distribute, and administer medications in a hospital has increased. A case study is performed that justifies the purchase of ADCs; primarily due to reduced nursing time in the travel and administration of medications.

Pazour and Meller [20] study the problem of locating medications in an ADC to minimize human errors and formulate this problem as a series of location problems. They develop heuristic methodologies that are computationally efficient and test their methodologies on a hospital example, testing situations with existing medications located in a drawer and when the drawer is empty.

5 Conclusions and Future Research

We analyzed the hospital pharmacy by drawing upon concepts from distribution center design. Our research focused on the similarities between a traditional distribution center and a hospital pharmacy, with special attention to order-fulfillment technologies. We reviewed the literature on analytical models for order-fulfillment technologies used throughout pharmaceutical distribution. A multitude of operations research techniques have been used to analyze the technologies including mixed-integer, linear programming, heuristic algorithms, probabilistic modeling, and simulation.

While order-fulfillment technology is used in both facilities, the driving force for implementation is different: in a distribution center, order-fulfillment technology is typically implemented to reduce lead times and labor costs. On the other hand, in a pharmacy, order-fulfillment technology is implemented to first enhance patient safety, and second to improve the pharmacy or nursing staff's productivity. The objectives used in the order-fulfillment technology models reflect this difference. Minimizing cost or time are common objectives for a traditional distribution center, while minimizing errors or maximizing quality are common objectives in the pharmacy-application domain.

The area of pharmacy system design presents a host of challenging problems. Future research could expand the analysis of functional processes beyond order-fulfillment within a hospital pharmacy. Because retail pharmacies tend to handle larger volumes of medications, retail pharmacy operations differ considerably from hospital pharmacy operations and could be analyzed. Also, because it is common for both a distribution center and a hospital pharmacy to use carousel systems, an analysis that addressed the unique nature of operating within a pharmacy would be interesting.

New initiatives are taking place in pharmaceutical distribution, including track-and-trace or pedigree programs. These programs are designed to help protect patients from counterfeit or tampered medications. The state of California has issued a mandate that all prescription drugs must have an electronic pedigree by 2015 and other states have begun to institute similar mandates. The California legislation requires that each prescription medication be electronically tracked at the item level through the distribution system, from the manufacture through distributors and wholesalers to its final transaction to a pharmacy. The implementation of this program presents a number of challenges and requires numerous changes to the current practices in the pharmaceutical supply chain. Interesting research questions specific to automation and distribution center design will be associated with these changes. For example, how will pedigree programs impact the ordering, handling, and distributing quantity? What are the logistic implications of tracking a unique product instead of bulk products through the pharmaceutical supply chain? What level of granularity is appropriate to track medications through the supply chain? What technologies should be employed to track medications, store medication information, and process information?

References

1. Aspden, P.: Preventing Medication Errors, Technical Report, Institute of Medicine of the National Academies, 500 Fifth Street, N.W., Lockbox 285, Washington, DC 2005 (2006)
2. Bartholdi, J.J., Hackman, S., Warehouse and Distribution Science, www.isye.gatech.edu/jjb/wh/, Release 0.87 (2008)
3. Bonsall, L., Heffner, S., 2007–2008 HDMA Factbook, Technical Report 0-9771914-8-6, Center for Healthcare Supply Chain Research (2007)
4. Bozer, Y.A., White, J.A.: A generalized design and performance analysis model for end-of-aisle order-picking systems. IIE Trans. **28**, 271–280 (1996)
5. Burns, L.R.: The health care value chain: producers, purchasers, and providers. Josey-Bass, San Francisco, CA (2002)
6. Caputo, A.C., Pelagagge, P.M.: Management criteria of automated order picking systems in high-rotation high-volume distribution centers. Ind. Manag. Data Syst. **106**, 1359–1383 (2006)
7. Center for Healthcare Supply Chain Research, 2009–2010 HDMA Factbook, Arlington, VA (2009)
8. Gu, J., Goetschalckx, M., McGinnis, L.F.: Research on warehouse operation: a comprehensive review. Eur. J. Oper. Res. **177**, 1–21 (2007)
9. Gu, J., Goetschalckx, M., McGinnis, L.F.: Research on warehouse design and performance evaluation: a comprehensive review. Eur. J. Oper. Res. **203**, 539–549 (2010)
10. Hassini, E.: Carousel storage systems. In: Lahmar, M. (ed.) Facility Logistics, Approaches and Solutions to Next Generation Challenges, pp. 199–234. Taylor & Francis Group, Boca Raton, FL (2008)
11. Heragu, S.S.: Facilities Design, 2nd edn. iUniverse, Lincoln, NE (2006)
12. Jernigan, S.: Multi-Tier Inventory Systems with Space Constraints, PhD Thesis, Georgia Institute of Technology (2004)
13. Kelle, P., Schneider, H., Wiley-Patton, S., Woosley, J.: Healthcare supply chain management. In: Badiru, A.B. (ed.) Inventory Management Non-Classical Views, pp. 99–127. Taylor & Francis Group, Boca Raton, FL (2009)

14. Liu, P., Zhou, C., Wu, Y., and Xu, N.: Slotting the Complex Automated Picking System in Tobacco Distribution Center. Proceedings of the IEEE International Conference on Automation and Logistics, Quingdao, China (2008)
15. Markowitz, A.J.: Making Health Care Safer: A Critical Analysis of Patient Safety Practices, Technical Report for the Agency for Healthcare Research and Quality, Contract No. 290-97-0013, University of California at San Francisco (UCSF)-Stanford University Evidence-based Practice Center, http://www.ahrq.gov/ CLINIC/PTSAFETY/index.html#toc (2001)
16. Mason, S.J., Thomas, L.M., Meller, R.D., Pazour, J.A., Root, S.E.: Survey of hospital pharmacy directors: assessment of the current state of unit-dose acquisition. J. Pharm. Tech. **26**, 3–8 (2010)
17. Meller, R.D., Klote, J.F.: A throughput model for carousel/VLM pods. IIE Trans. **36**, 725–741 (2004)
18. Meller, R.D., Pazour, J.A., Thomas, L.M., Root, S.E., Churchill, W.W.: The case for third-party repackaging in hospital pharmacy unit-dose acquisition. Am. J. Health Syst. Pharm. **67**, 1109–1114 (2010)
19. Park, B.C., Frazelle, E.H., White, J.A.: Buffer sizing models for end-of-aisle order picking systems. IIE Trans. Des. Manuf. **31**, 31–38 (1999)
20. Pazour, J.A., Meller, R.D., A Location Problem for Medications in Automated Dispensing Cabinets, Technical Report, University of Arkansas (2010)
21. Pazour, J.A., Meller, R.D.: Modeling the Inventory Requirements and Throughput Performance of Picking Machine Order-Fulfillment Technology, Technical Report, University of Arkansas (2010)
22. Pazour, J.A. Meller, R.D.: An analytical model for a-frame system design. IIE Trans. (in review)
23. Pazour, J.A., Root, S.E., Mason, S.J., Meller, R.D., Thomas, L.M.: Selecting and allocating repackaging technology for unit-dose medications in hospital pharmacies. Int J Innov. Technol. Manag. (in review)
24. Pederson, C., Schneider, P., Scheckelhoff, D.: ASHP national survey of pharmacy practice in hospital settings: dispensing and administration 2008. Am. J. Health Syst. Pharm. **66**, 926–946 (2009)
25. Perry, R.F., Hoover, S.V., Freeman, D.R.: An optimum-seeking approach to the design of automated storage/retrieval systems. Proc Winter Simul. Conf. 348–354 (1984), IEEE Press
26. Raghunath, S., Perry, R.F., Cullinane, T.: Interactive simulation modeling of automated storage/retrieval systems. Proc. Winter Simul. Conf. 613–620 (1986), ACM
27. Rules of Thumb, Warehousing and Distribution Guidelines. Technical Report 10th edn. TranSystem and Gross & Associates (2008)
28. VanDerLande Industries, Automated Order Picking/Order Fulfillment Systems, www.vanderlande.nl/web/Distribution/Products-and-Solutions/Order-Picking.htm (retrieved December 10, 2009)
29. Wise, L.C., Bostrom, J., Crosier, J.A., White, S., Caldwell, R.: Cost-benefit analysis of an automated medication system. Nurs. Econ. **14**, 224–231 (1996)
30. Yaohua, W., Yigong, Z.: Order-picking optimization for automated picking system with parallel dispensers. Chinese J. Mech. Eng. **21**, 25–29 (2008)

Automatic Scheduling of Nurses: What Does It Take in Practice?

Elina Rönnberg, Torbjörn Larsson, and Ann Bertilsson

1 Introduction

Many hospital wards need to be staffed by nurses around the clock every day of the week, and because of that, many nurses have to work irregular hours and according to schedules that have a great impact on their personal lives. Today there is a shortage of nurses in many countries, and in order to make the nursing profession more popular and to ensure high quality health care delivery, it is urgent to try to improve the working conditions for nurses. One possible and already ongoing improvement is that more flexibility and adaptation to personal requests is introduced in the scheduling.

Many nursing wards still use a traditional kind of scheduling, where the head nurse is responsible for creating a schedule by hand. Using this scheduling strategy leaves little room for flexibility and adaptation to personal requests from the nurses. Creating schedules by hand, or with limited computer support, is manageable, but both difficult and time consuming. The consequence of using this kind of scheduling strategy is that a lot of time and effort is spent on schedules that are neither favourable for the nurses nor satisfactory for the running of the ward. To instead use of a more modern approach for scheduling, of which self-scheduling is a common example, is an appealing alternative, but because of the increase in complexity that this entails, this kind of scheduling has proved to be difficult to use in practice.

E. Rönnberg (✉) • T. Larsson
Department of Mathematics, Linköping University, Linköping SE-581 83, Sweden
e-mail: elron@mai.liu.se

A. Bertilsson
Department of Mathematics, Linköping University, Linköping SE-581 83, Sweden

SCHEMAGI AB, Mjärdevi Science Park, Teknikringen 7, Linköping SE-583 30, Sweden

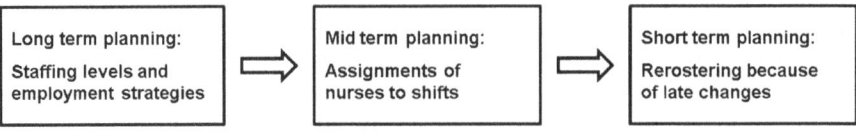

Fig. 1 The three phases of nurse scheduling

The nature of the staff scheduling problems, with their typically huge numbers of possibilities for how to construct schedules, generally make them suitable for addressing with operations research methods. Furthermore, if staff is needed round the clock, as is often the case for nurses, the staff scheduling problem provides a true challenge from an operations research point of view. Ever since the fifties, operations research techniques have been applied to nurse scheduling problems, but as Burke et al. [1] state in their article The state of the art of nurse rostering, Sect. 4, even though a lot of research has been devoted to applying operations research techniques to nurse scheduling, there is still a gap between the mathematical models and solution methods found in articles and the flexibility that is needed to tackle real-life nurse scheduling problems.

The intention of this paper is to provide a piece of practical experience that can help bridge the gap between advanced method development and the use of automatic nurse scheduling in practice. Our approach is to take on the real-life problem with all its details and to use a straightforward meta-heuristic in order to deliver automatically generated schedules. The contribution of this paper is based on the result of two case studies, which will provide insights into real-world examples, including evaluation and feedback from the wards.

1.1 Background

This section will provide a brief overview of the nurse scheduling problem all the way from the strategic planning, which is done well in advance of the scheduling, to the last minutes updates made on a daily basis. Some references will be given, but as a general starting point for further reading, the surveys by Burke et al. [1] and Cheang et al. [2] are recommended.

1.1.1 The Nurse Scheduling Process

The process of ensuring that there are enough nurses present at all times comprises of numerous decisions based on different time horizons and different levels of details. These decisions can be divided into three planning phases, as illustrated in Fig. 1.

The long-term planning is a part of the overall strategic planning process of each ward. First the ward managers must estimate how many nurses with each of the necessary skills are needed during all possible time periods of the day. Given these needs, the managers can determine how the days should be divided into working shifts and the staffing demand for each of these shifts. Based on this staffing demand, the kind of nurses necessary and their terms of employment can be decided. An example of such a condition is whether or not a nurse should work night shifts. The long-term planning is done for the first time when a new ward is opened, and the plan is then updated regularly, for example annually, and as well as whenever major changes occur. More information about this phase of the planning can be found in Arthur and James [3], Ghosh and Cruz [4], Venkataraman and Brusco [5], and Fagerström et al. [6].

When the staffing demand is known and there is a given workforce of nurses, each nurse is assigned to a schedule specifying which shifts she should work, usually for a scheduling period of 4–10 weeks. This phase in the planning process can be referred to as the mid-term planning, or nurse rostering. From an operations research point of view, this is the most well studied of the three phases of planning. Examples of surveys of staff rostering in general are Ernst et al. [7] and Ernst et al. [8], and surveys specialising in nurse rostering are Burke et al. [1] and Cheang et al. [2]. Mid-term planning is further described in the next section. The work presented in this paper mainly concerns mid-term planning, and this is henceforth what we refer to when the term scheduling is used.

The schedule produced during the mid-term planning should be seen as a viable plan, but the conditions for the staffing can change over time and necessitate some rescheduling. For example, a nurse can call in sick just before the shift starts, or the number of nurses needed can be different from the original estimate. Whenever there is a shortage of nurses for a shift, the short-term planning consists of deciding whether to use overtime, to call in a nurse on her day off, to call in a substitute nurse, or to try to manage despite the shortage. How to incorporate an optimisation tool in short-term planning is dealt with in Bard and Purnomo [9] and in Moz and Pato [10] for example.

1.1.2 Mid-Term Planning

There are two main approaches to creating a schedule for nurses; either the schedule can be unique for each scheduling period, or it can be cyclic, that is, the same schedule can be used repeatedly, period after period. In recent years, noncyclic scheduling has gained a lot of popularity due to the flexibility that follows from creating a unique schedule each period. Because both these approaches are common in practice, in our work we have chosen to present case studies in which one of the wards uses noncyclic schedules and the other uses cyclic ones. The details regarding the differences between cyclic and noncyclic scheduling will be presented in more detail in Sect. 2.5.

Whether the schedules are cyclic or not, there are some fundamental considerations that need to be taken into account when creating them. On the ward, there is a given staff of nurses that should be assigned to shifts. The nurses have different skills and individual contracts stating which shifts they can work and for how many hours per week they should work. For each shift, for example a day, an evening, or a night shift, there are given demands for nurses with certain skills, and these demands should be fulfilled.

In addition to assigning the nurses to shifts in such a way that the staffing demands are fulfilled, there are scheduling rules that prescribe the properties of a schedule. Firstly, the schedules must be consistent with the prevailing laws and regulations regarding staff scheduling. Secondly, there are always some quality aspects to consider for the schedule to be acceptable to the nurses and thereby possible to use. A typical example of such a quality aspect is the even distribution of unpopular shifts.

Given detailed information regarding the requirements described above, the task of creating a schedule for a nursing ward is complex, but doable. On many hospital wards, this work is still carried out manually, which is time-consuming and often results in schedules with undesired properties. Operations research methodology is well suited to addressing the mid-term planning problem, and a lot of research has been carried out within this field during the last five decades. Some examples can be found by Dowsland [11], Dowsland and Thompson [12], Bard and Purnomo [13], Berrada et al. [14], and Jaumard et al. [15].

As mentioned earlier, working a mixture of both weekdays and weekends, and both day- and night-time, has serious consequences on the personal lives of the nurses. The so-called preference scheduling is a popular way of trying to improve the working conditions for the nurses as it takes their requests and opinions into consideration when creating the schedule. However, the ambition to also take the nurses' desires into account further complicates the scheduling process.

A rather extreme form of preference scheduling is self-scheduling, which is a process in which nurses themselves are responsible for creating a schedule for each scheduling period. There are several ways of carrying out this process, for examples, see Bailyn et al. [16] and Karlsson [17]. Note however that even if the intentions behind using self-scheduling are good, it is often a real challenge to make it work well in practice when the schedule is created by hand.

1.2 Approach and Objective

Based on discussions with health care representatives who act on different levels of their organisations, we have concluded in earlier studies that a key to success for the automated scheduling of nurses is the ability to adapt the scheduling to the specific conditions on each ward. Based on this observation, we have developed a model designed to be easily adapted to different work places. To present all the details of this model is beyond the scope of this paper, so instead, we describe the kind of issues the model can handle, since this helps to answer the question of what it takes to automatically create a schedule for nurses in practice.

In order to carry out the case studies, we have designed a tabu search strategy focused on modularisation and flexibility rather than on short solution times. From a research point of view, this search strategy is of little interest, because it merely includes the standard components of a regular tabu search, see, for example, Burke and Kendall and Reeves [18]. For this reason, the tabu search strategy will be only briefly described in this paper, thereby allowing the focus to be on the results of two case studies carried out in order to demonstrate the wide variety of aspects to consider when scheduling nurses.

1.2.1 A Comment on the Choice of Solution Method

The reason for choosing a tabu search approach is twofold. Firstly, solving integer linear programs, such as scheduling problems, all the way to optimality is typically very difficult. Secondly, doing so in our setting is not practically meaningful, since when it comes to a schedule for nurses, the difference between an optimal solution and a good solution can be rather insignificant, because it might be just a few shifts that differ.

Why not use a branch and bound approach in a commercial solver and settle for a good solution there? Branch and bound is a viable approach, but a tabu search dedicated to the nurse scheduling problem is more likely to be more easily adaptable to different work places than a traditional integer programming model. Furthermore, when comparing these two solution processes and how the relaxation of the problem is made, there is a significant difference. In a traditional branch and bound approach, the integrality requirements are relaxed during the solution process, which means that the intermediate solutions do not constitute meaningful schedules. In our tabu search, the integrality is kept and instead some constraints are relaxed, which means the intermediate solutions during the search will always be complete schedules, though not necessarily fully feasible ones. In a nurse scheduling setting, the distinction between feasibility and infeasibility is not always clear, and rules are often expressed in terms of what is preferable, thus making it an advantage to easily be able to change which constraints should be relaxed and which should not and to be able to produce good and near-feasible solutions during the search.

1.2.2 The Case Studies

The two case studies to be presented have been carried out in three steps. First, we met with representatives of the ward in order to collect information about their ward and their scheduling. In the second step we ran the tabu search to find schedules to deliver to the ward, and during this step we had to ask the head nurse further questions about their scheduling conditions and suggest schedules in order to get feedback. The last step involved analysing and evaluating the final schedules.

As we will describe, the results of both the case studies show that we were able to automatically generate schedules that are at least as good as the manually constructed ones, which can be seen as a first step towards verifying that our approach is flexible enough for real-life nurse scheduling problems.

Due to practical reasons, both case studies have been carried out in Sweden, which is clearly a limitation. From what we know from the literature (see the

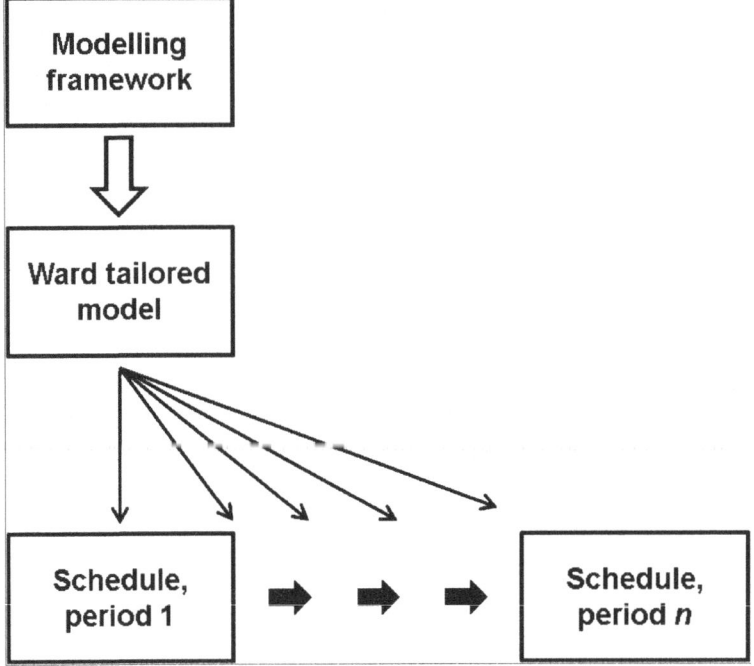

Fig. 2 Within the modelling framework, the model is tailored for the needs of each ward and then supplied with data for each scheduling period

references in Sect. 1.1) the aspects to consider do however seem quite similar in other countries, although some of them take slightly different forms.

1.3 Outline of the Paper

The two sections below will give a brief description of our mathematical model and tabu search approach. The two sections to follow then present the two case studies, and these are then followed by some concluding remarks.

2 Modelling Framework

As described in the introduction to this paper, in order to produce schedules applicable to a real-world nursing ward, it is crucial to be able to tailor the model to fit the rules and conditions on the ward in question. We do this in two steps, starting from a platform which we present in this section as our modelling framework. As can be seen in Fig. 2, the first step is to specify the characteristics of the ward in order to be able to create a model tailored to the ward, and then in a second step, for each scheduling period, include period-specific information, such as staffing demand and preferences from the nurses.

The overall design of the modelling framework is divided into the following blocks, which will be presented in the remainder of this section.

Minimise	The deviation from the staffing demand
Maximise	Fulfilment of requests from the nurses
	Fairness among the nurses
Subject to	Staffing demand
	Scheduling rules
	Quality aspects

In this section, the staffing demand, the scheduling rules, and the quality aspects will all be introduced as normal constraints. In reality some of these constraints are soft, which means that they only need to be fulfilled if it is possible. Which constraints are of this kind, and how important it actually is to fulfil them can vary from ward to ward. We ignore this distinction for the time being and will return to it when describing the tabu search in Sect. 3.

The aspects that need to be considered when automatically generating a schedule is heavily dependent upon the strategy the ward uses for their scheduling, and also on how their scheduling process is carried out. The modelling framework to be described could be used for cyclical as well as noncyclical scheduling and should preferably be used together with self-scheduling. The distinction between these different scheduling strategies and how they affect the automatic scheduling, will be addressed in Sect. 2.5.

2.1 Staffing Demand and Substitute Nurses

In the course of 1 day, a ward normally needs a varying number of nurses who possess different kinds of skills; something which should be matched to the members of a nursing staff, who typically have different education, training, and qualifications. In our model, the need for nurses is expressed in terms of a staffing demand, which is an estimate of how many nurses are needed during a shift. This staffing demand is accompanied by a profile describing the skills needed to fulfil the demand in question. For each shift, the staffing demand is given as a lower and an upper bound on the number of nurses needed.

Below follow some examples of differences between nurses on a ward.

- The nurses have different levels of education which place them in different categories, such as trained nurses and assistant nurses.
- There can be certain tasks that only some of the nurses have the proper training for.
- The nurses can be divided into groups, within each category of nurses or across the categories. The staffing demand may, for example, include that at least one nurse from each group covers a certain shift.

- Two wards partly share a nursing staff, meaning that some of the nurses, but perhaps not all, can cover shifts on either of the two wards.

When tailoring the model to a new ward, one specifies the kinds of categories, groups, and skills that are of relevance and what profiles are needed to cover a certain staffing demand. When it comes to the shifts, these are labelled to be either a day, an evening or a night shift, and it can be determined individually for each nurse when the shifts starts and ends.

If the staffing demand cannot be met by the nurses available, substitute nurses might need to fill the shortage, something which in practice costs money. For this reason, the use of substitute nurses should be kept at a minimum, hence making it an objective to minimise the number of under-staffed shifts. In addition, the number of over-staffed shift is minimised, because it is better to fill other not yet fully staffed shifts instead.

When creating the schedule, most of the staffing demand can be fulfilled in numerous ways by different combinations of nurses, but there is also a kind of staffing demand which usually concerns a predetermined nurse and shift, and because of that, the fulfilment of this demand is predetermined. Examples of such demands are holidays, administrative tasks, and meetings or courses.

2.2 Scheduling Rules

An automatically generated schedule must of course be consistent with rules such as prevailing laws and regulations regarding staff scheduling, and our means for accomplishing this will be presented here. These types of rules are mainly of the following two kinds.

- Rules of a combinatorial nature, stating which shifts can be combined and not.
- Rules concerning the number of hours worked.

The combinatorial constraints are easily handled by stating which combinations of shifts are possible and not, either for all nurses on a ward simultaneously or, if the rules differ among the nurse, for each nurse individually. Rules concerning how many days in succession that a nurse is allowed to work are handled similarly, and these rules can also be made dependent on the kind of shifts worked.

When it comes to keeping track of the number of hours worked, every nurse is assigned a nominal number of hours to be worked per week, and this is the number of hours she should work on average according to her employment contract. This nominal number of hours typically varies among nurses. In practice, it is not possible to allow the nurses to work their nominal number of hours every week, and a limited deviation must be allowed. This deviation is called a time status, and it is updated every week. Its bounds control the accumulated maximum deviation from the nominal number of hours.

A ward may wish to further control the deviation from the nominal number of hours to be worked. For this purpose, the model offers the possibility to control the number of hours each nurse works both during the whole scheduling period and during a single week, which can be set to start on a Monday or on any other day of the week. These hours are then allowed to vary within certain intervals which are set for each workplace. Combining these three measures correctly is a good means for evening out the workload for each nurse during a scheduling period.

2.3 Quality Aspects

In order to create a schedule that complies with the laws and regulations, it is sufficient to take into account the aspects of the scheduling described previously in this section, but considering these aspects only will not be sufficient to produce a schedule that can actually be used by a nursing ward. Below, we present the remaining considerations to be taken into account when creating a schedule; these will be referred to as quality aspects. At first glance, one might believe that quality aspects are less important than the rules described earlier, but this is actually not the case in practise. The quality aspects have shown to be just as important as the scheduling rules.

Legislation on working hours puts some limitations on how many days in succession a nurse is allowed to be at work, but on many wards there are additional rules whose purpose is to improve the working conditions for the nurses. A typical example thereof is a rule for how many nights in a row a nurse is allowed to work. These kinds of rules can be individual for each nurse, depending for example on her preferences and ability to work nights.

All the previously described rules for how to combine shifts have considered which shifts not to combine. It is also quite common that wards have policies for shifts that should be combined. A typical such example is that at least one of the nurses working the evening shift should also work the day shift the morning after, because then at least one of the nurses then is up-to-date about what happened the day before. In addition to the regular scheduling rules, a ward can also have a policy with additional rules on which shifts not to combine.

For natural reasons, some kinds of shifts, for example evening, night, and weekend shifts, are less popular among the nurses. These unpopular shifts should therefore, in some sense, be fairly distributed among the nurses. What is considered to be fair differs among wards, but it is crucial to distribute these kinds of shifts according to what is accepted on the particular ward. Some of the examples we have come across for night shifts are

- All nurses shall work the same number of night shifts.
- For each nurse, the number of night shifts worked shall be proportional to the percentage of full-time she works.

- Each nurse is assigned to a category, depending on the percentage of full-time she works, and for each such category it is determined how many night shifts shall be worked.
- The head nurse determines for each nurse individually how many night shifts she shall work.

Fairness can be desired within a group of nurses only or within the whole nursing staff. What also differs between wards is how they distribute the weekend shifts among the nurses. Some wards focus on the number of weekend shifts worked, while others only let the nurses work complete weekends, including at least shifts on both Saturday and Sunday, and therefore count whole weekends.

It is generally not possible to achieve complete fairness among the nurses, and therefore we keep track of the deviation between the number of an unpopular kind of shift that the nurse should work and the number that she has actually been assigned to. For some wards, it is crucial to achieve the best possible level of fairness, even at the expense of other preferences (this is typical for cyclical scheduling), while for other wards it is less important, as long as deviations are kept born in mind for the next scheduling periods and made up for in the long run.

2.4 Requests and Fairness

One important benefit of using operations research methods for scheduling nurses is how easily the nurses' preferences can be taken into account when creating the schedule. Some preferences come in the form of individually adjusted scheduling rules or quality aspects as described above, but most of them come as requests for when to work and not, and this kind of requests is in turn divided into two types, hard and soft. When creating our model, we introduced a system for making and handling these requests, and this system is presented here.

A hard request is directly translated into a staffing demand for the specific nurse and shift, and such a request must be approved beforehand by the head nurse. Examples of such requests are holidays, courses, and meetings. Some wards also allow for hard requests based purely on personal preferences, as for example to be free on a Wednesday evening because you want to go to the theatre. How and when the nurses can make hard requests is regulated by a policy created by the head nurse.

A soft request gives the nurse the possibility to have preferences about when to work and not, and this kind of request will be fulfilled if possible. The nurse can grade the request to be standard or important. When introducing the possibility to make requests, one challenge is to measure the fulfilment of requests in a fair way, including taking into account that

- There are different types of soft requests.
- Nurses working different percentages of full-time work different numbers of shifts, which affects the number of requests that can be fulfilled.

- Nurses with many hard requests, for example those having a weeks holiday, have fewer days for which to make soft requests, and this affects the number of requests that can be fulfilled.
- Each nurse has her own strategy for making requests, using many or few, or perhaps even trying to cheat the system.

The level of request fulfilment must be measured in a fair way, without leaving the nurses a possibility to take unfair advantage of the system. The main principles for how this is accomplished is described in the article Rönnberg and Larsson [19]. The calculations presented there lead to a normalisation of the fulfilment of requests, and the final outcome of these calculations is referred to as a score for each nurse.

One objective when creating the schedule is to maximise the total sum of scores, but if this objective is used alone, the scores of the resulting schedule are likely to vary significantly among the nurses, and therefore the schedule will probably not be acceptable to them. Fairness is the key to achieve acceptance, and therefore an additional objective is to maximise the score for the least favoured nurses. Furthermore, a high fulfilment of requests and a high level of fairness in every scheduling period is difficult to achieve without violating scheduling rules or allowing the schedule to have other unwanted properties. It is therefore vital to find the right balance between the objectives and also to bear the score in mind between scheduling periods and strive for complete fairness in the long run.

2.5 Scheduling Strategies

Depending on the context in which the schedule is to be used and on how the scheduling process is carried out, what is considered to be a good schedule and not can differ. The two main categories of scheduling strategies are cyclical and noncyclical, which will both be described below. A third strategy, which will also be commented on, is self-scheduling, a kind of noncyclical scheduling that has quickly grown in popularity during the past two decades. Our modelling framework is designed such that all of these strategies can be applied.

2.5.1 Cyclical Schedules

Using a traditional cyclical schedule means repeating the same schedule over and over again until the ward decides to change the schedule. Each schedule is typically of a period of between 4 and 10 weeks and is used for 6–12 months. Since the same schedule is used repeatedly it is very important that it is almost perfectly fair with respect to the score, the distribution of unpopular shifts, the number of hours worked, and quality aspects. This choice of scheduling strategy imposes the boundary restriction that the first week of the schedule can follow on from the last week. Further, because the nurses are bound to use the same schedule over and over again, they typically have a low level of influence on the scheduling when this kind of strategy is used.

2.5.2 Noncyclical Schedules

Noncyclical scheduling means that a new schedule, usually for a period of 4–10 weeks, is created for each scheduling period. The advantage of creating a new schedule for each period is that it offers greater flexibility due to the possibility to take into account both changes on the ward and period-specific requests from the nurses. A boundary condition is that the first weeks schedule is affected by the last weeks schedule from the previous scheduling period.

Instead of making the schedule very fair in each period, it can be made reasonably fair and then information about for example the score, the distribution of unpopular shifts, and the number of hours worked can be passed on to next period, and in this way a high level of fairness can be achieved in the long run.

2.5.3 Self-Scheduling

Self-scheduling is a general term used for the kind of scheduling processes where the nursing staff is jointly responsible for creating the schedule. This kind of scheduling exists in different forms around the world, but here our description is restricted to a kind of self-scheduling used in Sweden. Described briefly, this Swedish self-scheduling involves the following steps.

1. Without taking into account the staffing demand and other nurses' preferences, each nurse individually proposes a schedule for herself.
2. An improved and more feasible schedule is created through informal negotiations between the nurses.
3. A scheduling group consisting of around four nurses makes further improvements to the schedule.
4. The head nurse makes some final adjustments and approves the schedule.

Through contacts with several Swedish nursing wards, we have learnt that this kind of self-scheduling is much appreciated by the nurses, but unfortunately also associated with considerable problems. Making adjustments through informal negotiations often turns into a long and cumbersome process, which makes this kind of self-scheduling not only very time-consuming but also a source of conflicts. These problems however are not unique for this kind of self-scheduling, but are instead quite common for self-scheduling in general, see Bailyn et al. [16].

One great advantage of self-scheduling is that the scheduling is individualised and that the nurses can influence on when they work. The value of this influence is the reason why many Swedish wards use self-scheduling, despite the drawbacks presented. For a complete description of this kind of self-scheduling, the reader is referred to Rönnberg and Larsson [19].

3 A Sketch of the Tabu Search Design

The intention of this section is to provide a brief overview of how the tabu search algorithm explores the set of solutions on its quest for finding a solution good enough to be chosen as a schedule to be delivered to a ward. The main considerations in a tabu search, such as constraint relaxation, choice of neighbourhood, the evaluation function, and the search strategy, will be briefly commented on below.

3.1 Constraint Relaxation and Neighbourhoods

As described in Sect. 2, the model is very flexible with respect to the constraints it is possible to include. It is common that some of these constraints are more important than others. Some constraints must always be complied with, while others serve more as guidelines and preferences. Because of this distinction, the constraints are divided into two categories, hard and soft, where a hard constraint has to be complied with, and a soft constraint is to be respected if possible. The constraints of the latter kind can in turn be ranked according to their importance. A common example of how this categorisation could be applied is to forbid a nurse to work 6 days in a row, while working 5 days in a row can be set to be unwanted but still feasible.

During the search, each intermediate solution is a schedule where the nurses have been assigned to shifts. On the one hand, to allow the search to explore feasible solutions only can prevent the search from reaching areas in the solutions space that can be of interest, and therefore some of the hard constraints are relaxed and the violations of these constraints are instead penalised in the evaluation function. On the other hand, in order to preserve the structure of the problem and also to keep the neighbourhoods within a reasonable size, some of the hard constraints are always required to be fulfilled during the search. As a consequence, the only conditions for a schedule to be a possible intermediate solution during the search is that all non-relaxed hard constraints and all hard requests are satisfied.

In general, a move is made by changing the shifts to be worked by a nurse, and all possible such moves constitute the neighbourhood. The neighbourhood is changed during the search, and sometimes only a subset of the full neighbourhood is explored. Such a subset is chosen primarily as it is considered to be promising with respect to the current phase and also with some randomness included. In order to avoid discrimination if a tie occurs, it is broken by making a random choice. The strategy of including randomness in some decisions also helps to diversify the search and to avoid cycles.

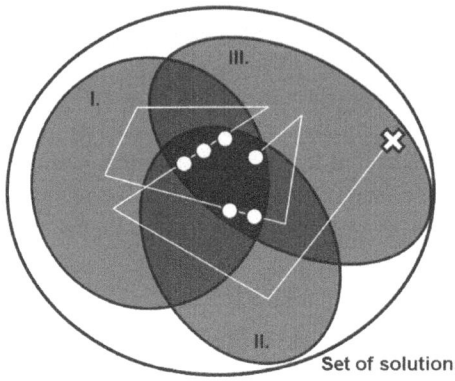

Fig. 3 The *white line* illustrates how the set of solutions is explored and a sharp turn represents a change from one phase to another

3.2 Evaluation Function and Search Strategy

The three original objective functions introduced in Sect. 2 were to minimise the deviation from the staffing demand, to maximise the fulfilment of requests, and to maximise fairness among the nurses. These three objectives, together with the penalties for violating soft constraints and relaxed hard constraints, will constitute the evaluation function.

The different contributions to the evaluation function are grouped into three components in order to enable the search to be guided to focus on different aspects during different phases of the search. These three components are the following.

Staffing:	To reduce deviation from the staffing demand
Hard constraints:	To reduce violation of the relaxed hard constraints
Preferences:	To reduce violation of soft constraints, to improve the fulfilment of requests, and to achieve fairness

Each component is represented by a function obtained by weighing together the contributions to be taken into account in that component. The components are then weighed together to obtain the evaluations function.

The search starts from a solution, that is a schedule that fulfils the non-relaxed hard constraints and all requests made by the nurses. From the point of view of each individual nurse, this initial solution is an optimal one, but it does not constitute a complete feasible schedule. The aim of the tabu search is therefore to find a good and feasible solution which is as close as possible to the initial one. The change of focuses among the three components is obtained by changing the weights between the components, yielding the behaviour known as strategic oscillation. The phases changes are illustrated in Fig. 3, where a sharp turn in the search represents a change of focus.

Changing focus during the search also serves as diversification, while keeping the same focus for a long time serves as intensification. When to change phases and what to focus on is controlled by factors such as the number of iterations within the same phase and the results of the moves during the phase, as well as randomly triggered events.

3.3 Tabu Lists, Aspiration, and Termination

Static as well as dynamic tabu lists are used in order to avoid cycling. Some of these are effective within a phase, while others are designed to control what happens between phases. The tenures of the lists are dependent on the size of the scheduling problem under consideration.

As an aspiration criterion, all tabu lists are ignored if a move leads to a solution that violates no hard constraints, has no deviation from the staffing demand, and has the best evaluation function value ever. The search terminates after a predetermined number of moves or when a predetermined number of feasible schedules has been found.

3.4 Outcome of the Search

As will be demonstrated by the case studies in the next section, there can be significant differences between wards. One consequence of these differences is that different measures are of interest when evaluating outcomes of the automatic scheduling. What measures are of interest is considered both when tailoring the model to the ward and also when setting the parameters for the tabu search. Examples of such parameters are the weights in the evaluation function and the tenures of the tabu lists. Making the right choices of parameter settings or not, can make the difference between success and failure. This lack of stability with respect to the parameter settings is of course a drawback of the solution method used, and also a reason for continued research to find a method less dependent on parameter settings.

4 Case Study 1

The first case study was carried out on a hospital ward specialised in treating patients with infectious diseases. This case is a continuation of the pilot study presented by Rönnberg and Larsson [19], which was carried out on this same ward. Within the scope of the previous pilot study, we used an AMPL (see [20]) implementation of a predecessor of our current model and CPLEX (see [21]) as solver for delivering schedules to the ward. During the long collaboration that we have had with this ward, we have delivered CPLEX-generated schedules which they have used. With regard to the three schedules presented in this section, these were

Table 1 The number of nurses working a certain percentage of full-time

Part of full-time	60%	75%	80%	85%	90%	100%
Trained nurses	0	4	2	1	1	8
Assistant nurses	1	6	1	4	0	2

the first ones created using our tabu search method, and they were therefore not used, but only evaluated. After these three periods, the collaboration continued and we have delivered a schedule for the ward to use, instead of them making one themselves.

Since all the schedules delivered so far have had the same characteristics, we find it sufficient to present the first three only. Moreover, during these three periods, the model remained almost unchanged, and because of that, some data will only be presented for one of the periods, to serve as an example.

The scheduling process used on this ward is self-scheduling, a brief description of which is found in Sect. 2.5 and a more thorough one in Rönnberg and Larsson [19]. Because the nurses on this ward are used to self-scheduling, the fulfilment of requests for shifts is of great importance to them, and therefore the analysis of the result will focus on the fulfilment of requests. As will be presented, the ward uses no soft constraints but only hard ones.

4.1 The Ward

The ward has the capacity to care for 30 patients at a time, and it is staffed around the clock. The staff usually consists of about 15 trained nurses and 15 assistant nurses (in Swedish: sjuksköterskor and undersköterskor, respectively), with the difference between the two categories being that the trained nurses have a university degree in nursing, while the assistant nurses have only studied nursing at an upper secondary school. The trained nurses have more responsibilities and are permitted to perform more qualified tasks than the assistant nurses. The nursing ward is integrated with an open clinic, also specialised in infectious diseases, and some of the trained nurses work both on the ward and in the clinic.

The ward uses three types of shifts, day (approximately 6.45 a.m.–3.15 p.m.), evening (approximately 1.30 p.m.–9.45 p.m.), and night (approximately 8.45 p.m.–7.00 a.m.). The times given are only approximate because there is a flexibility of ±30 min in when the shifts start and end, depending on the nurses' contracts. The open clinic is only staffed during the daytime, 8.00 a.m.–4.30 p.m.

The number of nurses employed on the ward varies somewhat, but in our example, taken from scheduling period 3, 16 trained nurses and 14 assistant nurses were working on the ward. Most of these nurses, namely 12 trained nurses and 13 assistant nurses, work all the three types of shifts, while the other nurses work day and evening shifts only. Not all of the nurses work full-time, and the percentage of full-time they work can be seen in Table 1. For trained nurses, full-time means 37 h per

Table 2 The staffing demand for trained nurses/assistant nurses/total number of nurses

	Mon	Tue	Wed	Thu	Fri	Sat	Sun
Day, min	3/1/7	3/1/7	3/1/8	3/1/7	3/1/7	2/1/5	2/1/5
Day, max	6/4/7	6/4/7	7/5/8	6/4/7	6/4/7	4/3/5	4/3/5
Evening, min	2/1/5	2/1/5	2/1/5	2/1/5	2/1/5	2/1/4	2/1/4
Evening, max	4/3/5	4/3/5	4/3/5	4/3/5	4/3/5	3/2/4	3/2/4
Night, min	1/1/3	1/1/3	1/1/3	1/1/3	1/1/3	1/1/3	1/1/3
Night, max	2/2/3	2/2/3	2/2/3	2/2/3	2/2/3	2/2/3	2/2/3

week, if they work day and evening shifts only, and 36.2 h if 12 they work all kinds of shifts. For the assistant nurses, the corresponding number of hours are 38.15 and 36.2, respectively.

For each shift during a week, the staffing demand specifies the minimum and the maximum need for trained nurses and assistant nurses, as well as the total number of nurses needed. Because the staffing demand was the same during each scheduling period and almost the same for all periods, we give only an example of a typical week in Table 2.

The staffing demand for the open clinic usually requires need one trained nurse each day shift during weekdays. This demand can be fulfilled by a few of the trained nurses only, and the shifts at the clinic should be distributed evenly among these nurses.

Both night and weekend shifts should be evenly distributed among the nurses. However, if there would be a nurse working 50% of full-time or less, she should work only half as many shifts on both nights and weekends, as a nurse working full-time.

The combinatorial kinds of constraints used are

- A nurse can work at most 5 days in succession.
- A nurse can be scheduled to work at most three consecutive night shifts.
- A nurse can work at most one shift per day.
- After working a night shift, a nurse may not be scheduled to work on the day or the evening shift the following day, or the day shift on the day after that.
- When working during a weekend, a nurse works either only days and evenings, or only nights.
- If working a weekend comprising of days and evenings, the nurse shall work both Saturday and Sunday, as well as either the evening shift on Friday or the day shift on Monday.
- There are no special rules for working nights over a weekend.

The number of hours worked is controlled by a time bank and intervals for the number of hours worked, both per week and in total during the period. The time bank is set to ±20 h and the number of hours worked is allowed to vary between 50% and 160% of the nominal values within a week, from Monday to Monday and from Friday to Friday, and with ±20% within the period.

Table 3 The number of requests made by the nurses

Type of request	Period 1	Period 2	Period 3
Hard request, holiday	7	54	37
Hard request, meetings, etc.	54	17	51
Hard request, not to work	144	375	369
Hard request, to work	0	40	9
Soft requests, important not to work	0	48	0
Soft requests, important to work	0	25	0
Soft requests, to work	494	807	836

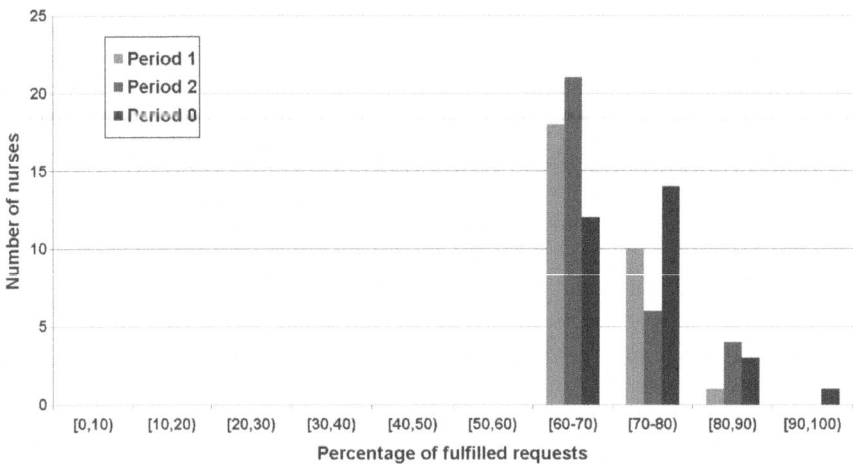

Fig. 4 A histogram showing how many nurses that for each scheduling period got a certain number of requests fulfilled

4.2 The Schedules

We will present the results from three scheduling periods of, respectively, 5, 8, and 8 weeks. As stressed earlier, the fulfilment of their requests is of great importance to the nurses on the ward. Table 3 shows the kinds of requests used for our scheduling on the ward, and how many requests of each kind that were made.

The solution times for the tabu search varied from a few hours to a few days. We find this acceptable since no major effort has been made to reduce these times, and also keeping in mind that a new schedule is needed only about five to ten times a year. The tabu search was able to find feasible schedules, both with respect to the staffing demand and the hard constraints, for all three scheduling periods.

An interesting measure for evaluating the outcome of the scheduling is to count the percentage of fulfilled requests each nurse receives. A histogram showing this result is presented in Fig. 4, and as can be seen there, fulfilment is both quite high and quite fair, with a request fulfilment rate of at least 60% for each nurse.

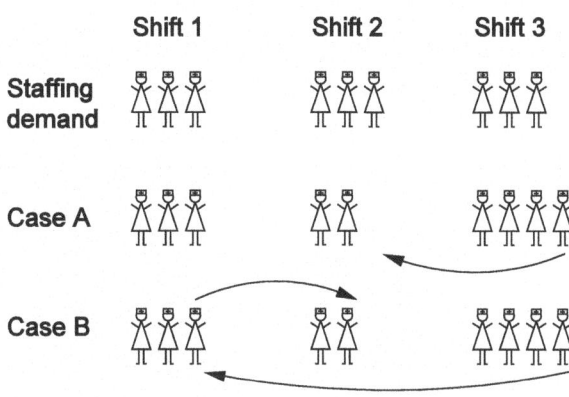

Fig. 5 Illustration of the two kinds of swaps; in Case A only a primary swap is carried out, and in Case B both a primary and a secondary swap are carried out

It is worth mentioning that the results of period three are from a second attempt by the tabu search to find a schedule for this period. When running the tabu search the first time, it was discovered that the input data was not consistent with respect to the number of nights requested and the number of nights that should be worked by some of the nurses. After adjusting these numbers, the tabu search obtained the result presented here.

The remainder of this section is devoted to evaluating and analysing viability of automating the self-scheduling process by using our tabu search method. A cornerstone in the Swedish self-scheduling process presented is that nurses shall request as many shifts as they should work. The deviation between the requests made and the final schedules can therefore be represented by swaps of shift; that is, if a nurse does not work a requested shift, then she has to work some other shift. The swaps can be divided into two types, primary swaps and secondary swaps.

By primary swaps is meant those swaps that are directly caused by the difference between the staffing demand and the requests made. The primary swaps are sufficient to redistribute nurses from over-staffed shifts to the ones with a shortage, but in order to create a schedule that is also consistent with the scheduling rules, further swaps might be required, and those are called secondary. The two types of swaps are illustrated in Fig. 5. In this example, there are three shifts that need to be staffed, which can represent any one of the three shifts in a scheduling period, and they do not need to be consecutive or ordered. The staffing demand is three nurses for each shift. In their proposed schedules, three nurses request the first shift, two request the second, and four request the third one. In Case A, it is possible to let one of the nurses who requested to work the third shift to work the second shift instead, which requires only a primary swap. In Case B, it is not possible for any of the nurses who requested the third shift to work the second one, for example because they do not have the right skills. Instead, one of these nurses is moved to shift one, which she is allowed to work, and then one of the nurses from shift one, who can work the second shift, is moved to the second shift. In this case, both a primary swap and a secondary swap are required.

In the analysis which follows, we only consider requests for working shifts. In Fig. 6 there are three bars that represent the result for each scheduling period. The

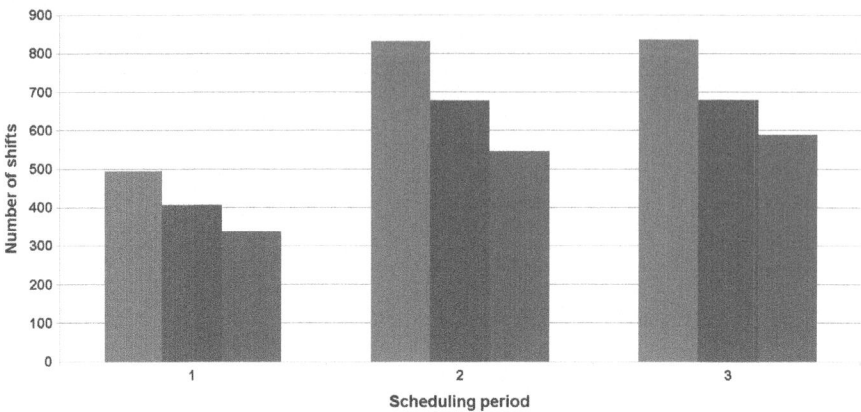

Fig. 6 Fulfilment of requests for working a shift. For each scheduling period the first bar shows the number of requests made, the second one how many of these coincide with the staffing demand, and the third one shows the number of requests fulfilled

first bar shows the total staffing demand for the period. The second bar shows the number of requests that coincide with the staffing demand, that is to say the difference between the first and the second bar is the number of primary swaps required. The last bar represents the number of requests for working, of any strength, that were fulfilled by the tabu search. Hence, the difference between the second bar and the third bar is the number of secondary swaps, plus the number of requests not fulfilled because the schedule does not correspond to an optimal solution.

In our example, the total staffing demand is nine nurses, and for each of the shifts, the number of requests that coincide with the staffing demand are three, two, and three, respectively. In the kind of diagram described above, the first and the second bar would be nine and eight units high, respectively, with the difference corresponding to the single primary swap required. In Case A, the third bar would be of the same height as the second one, since only a primary swap is needed, while in Case B, the third bar would be one unit lower, because of the secondary swap.

The number of primary swaps required can clearly not be affected by any optimisation tool, or by any other approach for creating the schedule, because it depends solely on the difference between the staffing demand and the requests made by the nurses. If this difference should be large, there would be no reason to use self-scheduling at all, because it would then not be possible to meet many of the requests. However, as is illustrated in Fig. 6, about 81% of the requests made in this case study coincide with the staffing demand; which we find very promising.

One welcome result is that there were as few secondary swaps as is shown in Fig. 6. Although the secondary swaps are induced by the scheduling rules, the number of such swaps is highly dependent on how the ward is staffed. If all nurses work a high percentage of full-time, then it is more difficult to create a schedule than if there are more nurses, who all work a lower percentage of full-time. The number of nurses who have the same skills and who therefore can fulfil the same demand is also important, since this affects the likelihood that a primary swap will be sufficient.

4.3 Feedback from the Ward and Discussion

The results presented show that the model and the method used were able to create feasible schedules for all three scheduling periods. The schedules were reasonably fair and all complied with at least 60% of the requests made by each nurse.

The head nurse seems convinced that it will be possible to automatically generate schedules for the kind of self-scheduling that they use, but that how the automatic generation is implemented and presented to the nurses will be crucial if it is to receive acceptance. She also agrees that it is important to consider the schedule created by this tool as a suggestion that can then be adjusted by the nurses themselves, if such adjustment leads to an improvement. The schedules presented were in need of some minor adjustments, something that probably is, and always will be, inevitable; but compared to the work of creating the schedules manually from scratch, this workload is negligible.

The fact that our collaboration with this ward continued after the case study and that we have since delivered a schedule that is to be used, is a result that we believe speaks for itself.

5 Case Study 2

The second case study was carried out at a home care unit in Norrköping, where they use a 6-week cyclical schedule, which is typically repeated for around 6 months. At this work place, the nurses are not allowed to make requests in a self-scheduling sense, due to the difficulties the head nurse has experienced in handling the requests when creating the schedule manually. Instead, the head nurse allows the nurses to make a very limited number of simple requests, such as not being scheduled on Tuesday evenings.

Instead of evaluating the schedule with respect to the fulfilment of requests, as we did in Case study 1, the measure of quality here is associated with the ability to comply with soft constraints and to do it in a fair manner. The importance of complying with soft constraints is a consequence of using cyclical schedules; since the same schedule should be repeated over and over again, certain properties of the schedule are preferred to others, and fairness with respect to these is crucial.

The collaboration with the ward was initiated around 6 months before this paper was written and we have therefore only had the time to create one schedule for them, alongside their original scheduling procedure. At the same as this paper was written, the tabu search method presented was used to deliver a schedule for the ward.

5.1 The Ward

The 29 nurses employed on the ward are all assistant nurses, and they are divided into groups depending on which geographical area they primarily serve. The groups,

Table 4 The number of nurses working a certain percentage of full-time

Part of full-time	50%	55%	60%	70%	75%	77.5%	80%	85%	90%	100%
Assistant nurses	2	1	1	1	3	2	2	1	2	14

Table 5 The staffing demand for Group 1/Group 2/Group 3/total number of nurses

	Mon	Tue	Wed	Thu	Fri	Sat	Sun
Day, total	3/3/1/14	3/3/1/14	3/3/1/14	3/3/1/14	3/3/1/14	1/1/1/5	1/1/1/5
Day, long	1/1/1/5	1/1/1/5	1/1/1/5	1/1/1/5	1/1/1/5	1/1/1/5	1/1/1/5
Evening	1/1/1/6	1/1/1/6	1/1/1/6	1/1/1/6	1/1/1/6	0/0/0/1	0/0/0/1
Split	0/0/0/0	0/0/0/0	0/0/0/0	0/0/0/0	0/0/0/0	0/0/0/5	0/0/0/5

Group 1, Group 2, and Group 3, consist of 12, 10, and 7 assistant nurses, respectively. Not all the nurses work full-time, and the percentage of full-time they work can be seen from Table 4.

The three types of shifts they use are day shifts, evening shifts, and split shifts, which means that a nurse works half a shift in the morning and half a shift in the evening. The length of a shift is determined individually for each nurse, depending on the percentage of full-time she works and also on whether it is a weekend or a weekday. A day shift starts at 7.00 a.m. or at 8.00 a.m. and the length of the shift is between 4 and 8 h. The evening shifts are also between 4 and 8 h long, and the split shifts are 9 or 10 h long.

The staffing demands for the evening and the split shifts are only expressed in terms of the minimum number of nurses needed and can be seen in Table 5. For the day shifts, there is a total minimum number of nurses needed, of whom some nurses must work a long day shift.

Three of the nurses have a fixed schedule created by the head nurse, and two other nurses each have a fixed schedule for half of the period because they have specific tasks to perform then. In each scheduling period, every nurse is required to attend two meetings, one with her group and one with the whole nursing staff. These meetings are each scheduled as a 2-h afternoon shift.

When it comes to the scheduling rules, some of them apply to all the nurses while some may apply only to some nurses. The constraints will be presented such that regular constraints are presented first, and the exceptions are given directly thereafter. The hard constraints of a combinatorial nature are the following.

- Each nurse must be consecutively off duty from Friday to Sunday or from Saturday to Monday, at least once during the scheduling period.
- Each nurse must be off duty at least 14 days during the scheduling period.
- A nurse should never work more than two consecutive evenings. Exception: There is a nurse who works evenings only.
- A nurse can work at most one shift per day.

- A nurse is not allowed to work a solitary day in between days on which she is off duty. Exception: The nurse working 50% of full-time is excluded from this rule.
- If a nurse works during a weekend, then she must work both on Saturday and Sunday. Exceptions: The nurses working 50% or 60% of full time, and some nurses with a fixed schedule.
- The evening shift on Fridays shall only be staffed with nurses working the weekend to follow.
- If a nurse works during a weekend, she has to be off duty on either Monday or Tuesday the following week.
- Around two-thirds of the nurses are allowed to work at most 4 consecutive days, the remaining one-third is allowed to work at most 5 consecutive days. Exception: One nurse is allowed to work at most 3 consecutive days.

The number of shifts scheduled per week can deviate from between 55% and 135% from a nurse's nominal number of shifts to be worked. The time bank, which in this case is measured in numbers of shifts, was set to ±3 shifts during the period. An objective for the whole scheduling period was to minimise the deviation between the nominal and the scheduled numbers of hours.

The head nurse wanted four soft constraints to be taken into consideration. Ideally, she would like all of these to be complied with, but from experience she knows that this is not possible. For future reference, when the result is to be analysed, these soft constraints are listed below.

- Five days in a row: The one-third of the nurses allowed to work 5 days in a row prefer not to do so, and therefore they have a soft constraint that they shall work at most 4 days in a row. Exception: One nurse likes to work 5 days in a row.
- Solitary day off: A nurse should preferably not be off duty on a solitary day between 2 days of work.
- Evening, then off duty: Ideally, a nurse shall not be off duty the day after she works an evening shift. Exception: One nurse prefers this.
- Two evenings in a row: If possible, the nurses should not be scheduled to work two evenings in a row. Exception: The nurse who work evenings only.

Only a few requests were made by the nurses, and an example of such a request is a nurse who wanted to be off duty on Tuesday evenings. Because the number of requests was so very small, the reasonable ones were all complied with, and they will not be discussed further. When evaluating the fairness, the requests were not of interest; instead fairness was associated with the number of unpopular shifts each nurse had to work and with the fulfilment of soft constraints. The head nurse decided that the unpopular shifts should be distributed with respect to the percentage of full-time a nurse works, and with some support from us, she then concluded that a fair distribution of these shift would correspond to what is presented in Table 6.

To achieve this distribution of shifts was a very important objective during the tabu search. As can be seen from the table, the head nurse had a special agreement with the nurse working 60% as well as with the nurse working 55%, who works evenings and weekends only. Nurses with a schedule fixed beforehand are not shown in this table.

Table 6 The number of unpopular shifts of each kind a nurse working a certain percentage of full-time should be assigned to

Part of full-time	50%	55%	60%	70%	75%	77.5%	80%	85%	90%	100%
Weekend shifts	3	6	2	4	4	4	6	6	6	6
Evening shifts	5	24	7	5–6	5–6	5–6	5–6	5–6	5–6	7–8
Split shifts	0	0	0	2	2	2	3	3	3	3
Friday evening shifts	1	2	1	1	1	1	1	1	1	1–2

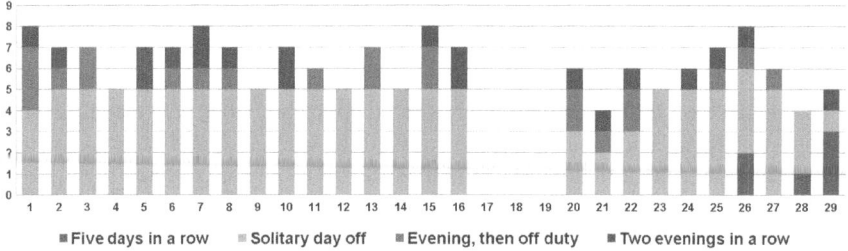

Fig. 7 For each nurse, the bar represents the number of times a soft constraint of each type has been violated in the schedule

5.2 The Schedules

For the analysis of this case study, the following four main issues are of interest.

- Does the schedule comply with the hard constraints?
- Is there a deviation between the staffing demand and the staffing levels?
- Is the distribution of unpopular shifts fair?
- How many times, if any, are the soft constraints violated for each nurse?

The tabu search was set to run for a few hours, and the outcome of the search was highly satisfactory, making it easy to answer the first three questions. The schedule complied with all the hard constraints as well as with the staffing demand. The distribution of unpopular shifts was made according to Table 6, which was a very welcome result since we did not know beforehand if this was possible. The deviation between the nominal and the scheduled numbers of hours was at most 3 h per nurse, something which is acceptable in practice. As was expected, some of the soft constraints were violated, and this will be further discussed.

The violations of soft constraints are presented in Fig. 7, with the number of occasions specified for each nurse. Because the constraints are not exactly the same for all the nurses, it could be rather difficult to draw any conclusions from the figures presented without taking the exceptions from the rules into consideration. In Fig. 7, the nurses are numbered from 1 to 29. The first 16 nurses are allowed to work at most 4 consecutive days, and they follow the regular constraints only. Nurses 17–29 are associated with some kinds of exceptions to be further discussed.

Nurses 17–19 have fixed schedules for the whole scheduling period, and they are therefore not affected by the soft constraints. For nurses 20 and 21, the schedule is fixed for one half of the scheduling period each, which means that they are less affected by the soft constraints, something which is also true for nurse 22, who also has a partly fixed schedule. Nurse 23 works evening shifts only, which means that some of the soft constraints do not apply to her. Number 24 is the nurse who wants to be off duty the day after she has worked an evening shift, and therefore the occurrences of such events are not included in the figure. Nurses 25–29 do not work many weekends, which makes it more difficult to avoid having them working 5 days in a row. Out of these five, nurse 29 prefers to work 5 consecutive days.

Studying the sixteen nurses following the regular soft constraints, one can conclude that it was difficult to avoid solitary days off, something which occurred four to five times for each of these nurses. The explanation to this is found when studying the other constraints and the staffing demand. The combination of having a schedule with both weekdays and weekends, along with the rules for how to work weekends (for example being free on Monday or Tuesday) makes it difficult to create many occurrences of consecutive days off, because the nurses have to alternate between working on weekdays and weekends. That each nurse has 4–5 solitary days off should be compared to the total number of days off for each nurse, which is at least 14.

The constraint preventing a nurse from having a day off after working an evening shift, as well as the constraint that forbids working two evenings in a row, are violated only a few times each, something which we find to be an acceptable result for a schedule to be usable in practice. Furthermore, from Fig. 7, it can be seen that the distribution of violated soft constraints is reasonably fair, and this is more important than the actual number of violations if the schedule is to be acceptable to the ward.

5.3 *Feedback from the Ward and Discussion*

Since the tabu search succeeded in complying with all the hard constraints, the staffing demand, and with the distribution of unpopular shifts, a lot of attention during this case study was given to the soft constraints. As described earlier, some of them were more easily complied with than others, and the ones to be prioritised was decided by the head nurse. During this study, we have focused on finding a best possible schedule for the ward in this particular scheduling period, and this has been done empirically, in close collaboration with the head nurse. If one were to further evaluate the behaviour of the method, it could be of interest to study the kinds of solutions that it is possible to create, and preferably perform tests on many instances from the same ward.

The nurses' responses varied between very satisfied with the result and a bit more negative because some soft constraints were violated. The head nurse's response to this criticism was that the soft constraints are also violated with manual scheduling, and that a great advantage of using our tabu search method is that she does not need to decide for which nurses and when this should happen. Furthermore, she also consider "the computer" to be more objective and fair in its decisions.

Usually the head nurse works for 1–2 weeks creating a schedule, and by using our automatically generated schedule, this time was reduced to 4 h. These 4 h were used for inspecting and evaluating the result of the scheduling and also for making small adjustments, such as extending or shortening shifts for the nurses in order to even out the deviation of ±3 h between the nominal and the scheduled numbers of hours for the period.

The most telling result of this case study is the fact that the ward wants us to deliver a schedule the next time they need a new one, and that we will do so.

6 Concluding Remarks

The long-term goal of our work is to develop an optimisation tool that functions as a practical tool for the automatic scheduling of nurses. That we managed to deliver usable schedules of high quality to both the wards in the case studies is very promising—and a huge leap in the right direction.

One objective of this paper was to illustrate the wide variety of issues to consider when scheduling nurses in practice. The two wards were chosen for the case studies because they differ significantly, and also because they are typical within their respective categories of Swedish nursing wards.

The tabu search method used does not play a central role in this paper, but instead merely serves as a means to be able to carry out the computations needed for the case studies. For future research within this area, it could interesting to develop a more efficient method, preferably less dependent on parameter settings, for solving the nurse scheduling problems described in our modelling framework.

Something that is common for all the wards that we have come across is the great importance of fairness, even though there are significant differences between the definitions of fairness. In Case study 1, fairness is associated with the fulfilment of requests only, while in Case study 2, it is associated with the soft constraints. It is also possible to combine the two aspects and let the fulfilment of requests and soft constraints interact. However, it can then be a challenge to obtain the right priorities between the two aspects.

One challenge when working with a new ward is to understand what is essential in their scheduling, and in order to be able to successfully deliver a schedule, it is of crucial importance to understand their values and traditions. During our work, the responses from the nurses have been both expectant and sceptical; expectant because of the time-consuming work and difficulties associated with the manual scheduling process, and sceptical mainly because they are afraid of losing control over the scheduling. Because of the nurses' scepticism, it is important to present the outcome of the automatic scheduling pedagogically and to emphasise that the optimization tool only offers a qualified suggestion for a schedule. If it is considered beneficial, the nurses are allowed to make minor adjustments themselves.

The great benefit that our approach brings should be the time and effort saved if the head nurse is handed a schedule that is both feasible and fair. Other benefits are the objectivity of a computerised planning tool and the decrease in lead time for constructing a schedule.

Acknowledgements Thanks to all the health care representatives that we have been in contact with for sharing your knowledge about nurse scheduling. Special thanks to Elisabet Shimekaw and Ann Andersson, head nurses on the wards in Case studies 1 and 2, respectively, and also to all the nurses working on these wards.

This work was carried out in collaboration between the Department of Mathematics, Linköping University, and SCHEMAGI AB, with financial support from HNV/Vinnova, and with support from LEAD.

References

1. Burke, E.K., de Causmaecker, P., van den Berghe, G., van Landeghem, H.: The state of the art of nurse rostering. J. Scheduling **7**, 441–499 (2004)
2. Cheang, B., Li, H., Lim, A., Rodrigues, B.: Nurse rostering problems—a bibliographic survey. Eur. J. Oper. Res. **151**, 447–460 (2003)
3. Arthur, T., James, N.: Determining nurse staffing levels: a critical review of the literature. J. Adv. Nurs. **19**, 558–565 (1994)
4. Ghosh, B., Cruz, G.: Nurse requirement planning: a computer-based model. J. Nurs. Manag. **13**, 363–371 (2005)
5. Venkataraman, R., Brusco, M.J.: An integrated analysis of nurse staffing and scheduling policies. Omega: Int. J. Manag. Sci. **24**, 57–71 (1996)
6. Fagerström, L., Rainio, A.-K., Rauhala, A., Nojonen, K.: Professional assessment of optimal nursing care intensity level: a new method for resource allocation as an alternative to classical time studies. Scand. J. Caring Sci. **14**, 97–104 (2000)
7. Ernst, A.T., Jiang, H., Krishnamoorthy, M., Owens, B., Sier, D.: An annotated bibliography of personnel scheduling and rostering. Ann. Oper. Res. **127**, 21–144 (2004)
8. Ernst, A.T., Jiang, H., Krishnamoorthy, M., Sier, D.: Staff scheduling and rostering: a review of applications, methods and models. Eur. J. Oper. Res. **153**, 3–27 (2004)
9. Bard, J.F., Purnomo, H.W.: Hospital-wide reactive scheduling of nurses with preference considerations. IIE Trans. **37**, 589–608 (2005)
10. Moz, M., Pato, M.V.: Solving the problem of rerostering nurse schedules with hard constraints: new multicommodity flow models. Ann. Oper. Res. **128**, 179–197 (2004)
11. Dowsland, K.A.: Nurse scheduling with tabu search and strategic oscillation. Eur. J. Oper. Res. **106**, 393–407 (1998)
12. Dowsland, K.A., Thompson, J.M.: Solving a nurse scheduling problem with knapsacks, networks and tabu search. J. Oper. Res. Soc. **51**, 825–833 (2000)
13. Bard, J.F., Purnomo, H.W.: Preference scheduling for nurses using column generation. Eur. J. Oper. Res. **164**, 510–534 (2005)
14. Berrada, I., Ferland, J.A., Michelon, P.: A multi-objective approach to nurse scheduling with both hard and soft constraints. Soc. Econ. Plann. Sci. **30**, 183–193 (1996)
15. Jaumard, B., Semet, F., Vovor, T.: Case study: a generalized linear programming model for nurse scheduling. Eur. J. Oper. Res. **107**, 1–18 (1998)
16. Bailyn, L., Collins, R., Song, Y.: Self-scheduling for hospital nurses: an attempt and its difficulties. J. Nurs. Manag. **15**, 72–77 (2007)

17. Karlsson P.: Rapport om kartläggning av arbetstidslösningar inom landstinget. Stockholms läns landsting (2005)
18. Reeves, C.R. (ed.): Modern heuristic techniques for combinatorial problems. McGraw-Hill, London (1995)
19. Rönnberg, E., Larsson, T.: Automating the self-scheduling process of nurses in Swedish healthcare: a pilot study. Health Care Manag. Sci. **13**, 35–53 (2010)
20. ILOG AMPL CPLEX System, Version 10.0, User's Guide. ILOG (2006)
21. ILOG CLEX 10.0, User's manual. ILOG (2006)

The manufacturer's authorised representative in the EU is Springer Nature Customer Service Centre GmbH, Europaplatz 3, 69115 Heidelberg, Germany. If you have any concerns regarding our products, please contact ProductSafety@springernature.com

Printed and bound by CPI Group (UK) Ltd, Croydon, CR0 4YY

23/03/2026

02076661-0002